SAFER SEAS
Systematic Accident Prevention

Koji Fukuoka

Manager, Occupational Health and Safety Section
Research Support Division

and

Emergency Management Specialist
Office of Chief Operating Officer
Okinawa Institute of Science & Technology Graduate University
Okinawa, Japan

CRC Press
Taylor & Francis Group
Boca Raton London New York

CRC Press is an imprint of the
Taylor & Francis Group, an **informa** business

A SCIENCE PUBLISHERS BOOK

i

CRC Press
Taylor & Francis Group
6000 Broken Sound Parkway NW, Suite 300
Boca Raton, FL 33487-2742

First issued in paperback 2021

© 2019 by Taylor & Francis Group, LLC
CRC Press is an imprint of Taylor & Francis Group, an Informa business

No claim to original U.S. Government works

Version Date: 20190515

ISBN-13: 978-0-367-77950-4 (pbk)
ISBN-13: 978-1-138-38893-2 (hbk)

Visit the Taylor & Francis Web site at
http://www.taylorandfrancis.com

and the CRC Press Web site at
http://www.crcpress.com

Preface

As a marine accident investigator at the Japan Transport Safety Board, the Ministry of Land, Infrastructure, Transport and Tourism (2009 to 2016), I used to conduct accident investigations and analysis. I was often asked by safety officers of the maritime industry and the shipping companies, why did accidents occur despite there being accident safety management systems in place. This induced me to study at the Graduate School of Kobe University (from 2013 to 2016) where I worked on accident occurrence mechanism, actual situation of occurrence of marine accidents, accident investigation method, accident analysis, accident model, accident prevention not only in maritime industry but also aircraft and nuclear power plants. I drew conclusions on accident models suitable for marine accidents, accident analysis methods and accident prevention measures.

My studies at the Graduate School coincided with significant improvements in the marine accident investigation field. From the beginning of the 21st century, accident investigation, analysis, and safety standards in the maritime industry, due to the efforts of the International Maritime Organization (IMO), have become more scientific using standard methods and techniques adopted by the IMO. These methods and techniques are discussed in this book.

The book is intended for those concerned with reduction in the number of accidents and incidents, safety officers in the maritime industry, marine accident investigators, researchers on accident prevention and those who are interested in safety. It explains the mechanisms of marine accident occurrence, accident investigation analysis method, and methods of preventing accidents and incidents scientifically in order from Chapter 2 to Chapter 11.

Chapter 1 covers the historical background on safety investigation in the maritime industry up until the scientific accident investigation was adopted by the IMO. Chapter 2 and Chapter 3 discuss the accident occurrence mechanism and accident models applicable to the maritime industry compared to other fields. Chapter 4 looks closely at the factors contributing of marine accidents, with actual accident cases obtained from the data of marine accident investigation reports. Chapter 5 readers to find the way of revealing defects in the safety management systems established their own companies. Chapters 6 and 7 mainly focus on the scientific accident investigation and analysis method recommended by the IMO, respectively. Chapter 8, after conducting the scientific accident investigation, illustrates the method to visualize defects in the system using the new accident model, Risk Management and Quality Management Process approach model (RMQMP model), which was developed by the Swiss cheese model. In addition, it clarifies the latent conditions for each type of accident, pattern of accident occurrence, and application and unresolved issues associated with the Swiss cheese model. Chapter 9 discusses quantification of accident occurrence using statistics derived from 89 cases of accident analysis and points out areas that organizations make more efforts to prevent accidents effectively and efficiently. Chapter 10 discusses the application of the RMQMP model to fields other than maritime industry in contrast to the Systems-Theoretic Accident Model and Processes (STAMP) model. Chapter 11 explains the systematic accident prevention: The process of utilizing the methods explained in the previous chapters to reduce the number of marine accidents. It also demonstrates, with detailed actual accident and incident cases, that the process of systematic accident prevention is applicable not only to accidents but also to incidents that did not lead to accidents. This chapter helps readers to see the defects in the system of their own organizations and the safety management systems, and to reduce the number of accidents and incidents by utilizing statistics.

Koji Fukuoka

Acknowledgements

I am indebted to many people who helped and encouraged me to study deeply into the prevention of accidents and to complete this book, without them this book would not have been possible. Although I cannot name all of them, I would like to thank those people whose assistance was indispensable. Prof. Masao Furusho of Graduate School of Maritime Sciences, Kobe University, provided me with valuable comments when I had been studying there. Prof. Graham Braithwaite and Nicholas Beer of Cranfield University, where I had learned about accident investigation, taught me a variety of investigation methods, analysis and accident models not only in the marine field but also aviation and train and human factors. Hideo Osuga, Secretary General of Japan Transport Safety Board, provided me with an invaluable opportunity to learn at Cranfield University. I must also express my great appreciation to Masaharu Okamoto, Representative of International Registries Fareast Limited Japan Branch, who had encouraged me to study at the Graduate School of Maritime Sciences, Kobe University. My sincere thanks to the staff of Japan Transport Safety Board, with whom I had often rushed to the accident site and investigated accidents across Japan, and the members of the Marine Accident Investigators' International Forum with whom I had discussed accident analysis and common approach to marine accident investigation.

I am especially indebted to my wife, Hiroko, and my daughters, Yuri and Reiko. They have supported me over the years while giving me cooperation and understanding during my studies and writing.

Contents

Chapter 1

History of Marine Accidents, Accident Investigation and Prevention

1.1 Prior to the loss of the *TITANIC*

During and after the industrial revolution in the United Kingdom in the 18th century, raw materials and manufactured goods were transported throughout the world mainly by ships. Seafarers had only the use of rudimentary navigational devices and charts. The accuracy of a ship's position mostly depended on circumstances such as weather, the waters being navigated, and the skills of navigators. There were no radars, gyro compass, global positioning systems (GPS), accurate navigational charts, significant navigational aids, predictable weather news, and seaworthy ships. Then there were no guarantees of a ship's safe return, but land-based society ignored the tragedies that happened out at sea. Maritime accidents increased according to the square of the increase in traffic by the middle of the 19th century in the UK. For instance, merchant navy was losing more than 2,000 ships a year. There were between 30,000 and 33,000 known wrecks around the coast of the UK.

In 1870, Samuel Plimsoll found that most shipwrecks were caused by overloading, and launched a parliamentary campaign on behalf of the British Merchant Seamen. He criticized shipowners' lack of concern for the safety of seafarers and the fact that they always stood to gain, irrespective of the outcome of the maritime adventure, and that what really mattered for them was the carriage of goods, and at worst, the collection of insurance money if their ships were lost.

Plimsoll's efforts materialized in the constitution of Merchant Shipping Act. The act granted strong power of inspection to the Board of Trade and introduced the Plimsoll line, which was intended to decrease accidents caused by overloading. But it wasn't until 1890 that the current Plimsoll line was established.

1.2 The *TITANIC*

In 1912, the loss of the *TITANIC*, regarded as an unsinkable ship, was significant both in scale of an accident and investigations that followed and were recorded and issued officially, both in the UK and in the United States.

The *TITANIC* sailed from Southampton, England, on her maiden voyage to New York at noon on 10 April 1912. She stopped at Cherbourg in France and Queenstown in Ireland, picking up additional passengers and mail, steaming west with 2,208 passengers and crew onboard. On 14 April, when she was proceeding about 700 miles east of Halifax, Nova Scotia in the North Atlantic at approximately 21.5 knots, one of the lookouts stationed at the crow's nest noticed an object in the distance, and then rang the warning bell, warning the bridge. However, it was too late for navigators to avoid a collision with the iceberg. Crew began preparing for evacuation, uncovering lifeboats. The capacity of lifeboats was little more than half of those onboard. Two hours and 40 minutes after the collision, the *TITANIC* broke into two and sank, taking more than 1,500 lives at 02:20 on 15 April.

According to the *TITANIC* Inquiry Project 2017, after the accident, two separate extensive investigations called inquiries were conducted in first the USA and then the UK.

The United State Senate inquiry was authorized and directed by the Committee on Commerce to summon witnesses to take testimony. The hearings were opened at the Waldorf-Astoria Hotel in New York city on 20 April 1912. Over the course of 18 days of inquiry, the testimonies of 86 witnesses were recorded, the inquiry transcript was over 1,000 pages long.

Since most of passengers and crew that survived were transported to the USA, the inquiry by British Wreck Commissioner on behalf of the British Board of Trade was conducted on 2 May 1912. The hearings were opened in the Wreck Commissioner's Court at the London Scottish Drill Hall on Buckingham Gate, and Caxton Hall on Westminster. The hearings closed on 3 July 1912.

1.3 SOLAS Convention

As a result of the sinking of the *TITANIC*, some 13 countries signed the first Safety of Life at Sea Convention (SOLAS Convention) in 1914 in order to regulate equipment and procedures intended to make seafarers safer. The regulations on SOLAS Convention included watertight bulkhead, fire-resistant bulkhead, lifesaving appliances, fire prevention, firefighting appliances, radio telegraphy, and the North Atlantic Ice Patrol. It did not include regulation of a marine accident investigation.

It wasn't until Inter-Governmental Maritime Consultative Organization (IMCO), which was established in 1948, adopted the SOLAS Convention 1948 that flag states were required to conduct investigations into incidents involving ships under their flags if the investigation might show regulatory issues as a contributory factor. The convention did not regulate the unified investigation methods and relevant regulations which harmonize the regulations that coastal states or flag states instituted. Therefore, each state conducted investigations under their own regulatory regime.

1.4 Accidents and international efforts

In the 1960s and 1970s, more and more maritime companies adopted flagging out, having their ships owned by other countries or Flag of Convenience (FOC), which became capable of the existence of open registry starting from 1948 by Liberian registry. As the number of FOC ships increased, the number of accidents within the sovereign water of a coastal state involving FOC ships was also increasing.

The *TORREY CANYON*, the Liberia flagged oil tanker, ran aground on Pollard rock on the Seven Stones reef, near Land's End, Cornwall, England due to mainly human error on 18 March 1967. The vessel was one of the first generation of supertankers. Thousands of tons of oil were soon spilling along the shores of the south coast of England and the Normandy coast of France from the ruptured tanks. During the next 12 days the entire cargo of approximately 119,000 tons of Kuwait crude oil was lost.

The *AMOCO CADIZ*, the Liberia flagged oil tanker, ran aground on Portsall rocks, three miles off the coast of Brittany, France, due to failure of the steering mechanism on 16 March 1978. The vessel had been en route from the Arabian gulf to Le Havre, France, when it encountered stormy weather which contributed to grounding. The resulting spill of 223,000 tons of crude oil polluted some 360 km of Brittany coastline.

In the wake of these marine accidents, the maritime industry acknowledged that loss of life at sea and marine pollution were unacceptable and had to be prevented, and that the gateway to the prevention of a marine casualty was an adequate investigation. International efforts lead by the IMCO were made in the form of resolutions in 1970s: The conduct of investigations into casualties, Exchange of information for investigations into marine casualties, and Personal and material resource needs of administrations for the investigation of casualties and contravention of conventions. During these periods, most coastal states investigated marine casualties under the framework of the United Nations Convention on the Law of the Sea (UNCLOS).

1.5 The *HERALD OF FREE ENTERPRISE* and *EXXON VALDEZ* disaster

The *HERALD OF FREE ENTERPRISE*, UK flagged roll on/roll off passenger and freight ferry, capsized about four minutes after sailing from Zeebrugge, Belgium, to Dover, England at 18:05 on 6 March 1987, manned by 80 crewmembers, laden with 131 vehicles, and 459 passengers in the good weather. Due to shallow water, the vessel came to rest with its starboard side above the surface. Water rapidly filled the vessel below the surface level. Not less than 150 passengers and 38 crewmembers lost their lives, and many others were injured. The position in which the vessel came to rest was less than seven cables from the harbor entrance. The extensive accident investigation found hardware issues, human errors and organizational issues (DOT 1987).

In this case, just like the *TITANIC* accident investigation, public enquiries were set up because a governmental organization specializing in marine accidents investigation had not been created in the UK. The board of enquiry which lead the public enquires put liability first and safety second, not leading to greater safety. In the aftermath of the capsize of the *HERALD OF FREE ENTERPRISE*, a new maritime accident investigation authority, Marine Accident Investigation Branch (MAIB), was established in 1989.

Similar to the UK, there was an increasing number of marine accident investigation organizations in the world: The USA, Canada, Australia, Sweden, France, etc.

The *EXXON VALDEZ*, the U.S. flagged oil tanker, ran aground on Bligh reef, Prince William Sound, Alaska in the U.S. at 00:09 on 24 March 1989. The resulting spill of an estimated 42 million liters of crude oil covered more than 26,000 km^2 of water in Prince William Sound and the Gulf of Alaska, and polluted more than 1,900 km of coastline. The accident extensive investigation conducted by the National Transportation Safety Board (NTSB) revealed, among other contributing factors, the importance of human factors, especially, effect of fatigue caused by overwork and lack of sleep (NTSB 1989).

1.6 Shift from hardware to human factors issues

As shown in the cases of the *HERALD OF FREE ENTERPRISE* and the *EXXON VALDEZ* marine accident investigations during the late 20th century, the focus of investigations transited from hardware issues to human factors and system issues. There are conflicting statistics among researchers on the contribution of human factors to transport accidents, but they all agree that human factors have become dominant in causes of accidents. Some experts estimated that it was about 20 percent in the 1960s, and about 80 percentage in the 1990s. Improved materials and engineering techniques in vessels, airplanes, and trains have brought the human factor into great prominence. Those statistics were realised by comprehensive accident investigations issued by an increasing number of permanent accident investigation authorities and by international efforts to set up a common approach for the investigations.

In marine accident investigations, the focus of transition has been partly influenced by accident investigations in aviation and nuclear power plant industries, and the development of human factors and accident models. These are closely related to the complexity of systems in socio-technical context in modern society.

In aviation, when radical automation of the flight deck was introduced by Airbus in the early 1970s, some of the emphasis on accident investigations had shifted to software issues, and a flight management system created human factor problems. The accident that became an epoch in aviation was the *Tenerife* disaster in 1977, recorded as the worst accident in aviation history, where two jumbo jets collided on the runway. The accident prompted the need to consider the standardization of communication among pilots and air traffic controllers, the need for assertion or proper power balance between a pilot and a co-pilot in a cockpit, etc.

In nuclear power plant industry, statics of human errors were abundant, and this data contributed to the development of different kinds of human factors and accident models by their own. As for the most influential accidents, there are three major accidents: The Three Mile Island disaster in 1979, which revealed managerial, maintenance system failures and error committed in control room, and the Chernobyl disaster in 1986, in which human errors and safety culture were highlighted. Finally, the Fukushima Daiichi nuclear power plant disaster after the attack of tsunamis triggered by earthquake of magnitude 9.0 off the coast of east Japan in 2013 has led people to reconsider the ability of emergency responses and resilience.

During 1970s to 1990s, when those major organizational accidents had been occurring across the fields, there were important developments in human factors and accident models. As one of the human factor models, the SHEL model which is abbreviation of software, hardware, environment, and liveware, was developed by Edwards in 1972, then renovated in the form of five blocks, and published (Hawkins 1984). The SHEL model was adopted as a tool to find human factors in aviation accidents by the International Civil Aviation Organization (ICAO). For an accident model, the Swiss cheese model (SCM) was developed (Reason 1990) and used for a background theory across the fields, including marine and aviation.

The International Maritime Organization (IMO: IMCO renamed in 1982) incorporated the SHEL model and the SCM in Resolution A.884(21) in 2000: Amendments to the code for the investigation of marine casualties and incidents—the IMO/ILO process for investigating human factors. Resolution A.884(21) was replaced by Resolution A.1075(28)—Guidelines to assist investigators in the implementation of the casualty investigation code in 2014; but the IMO model based on the SCM is used to help investigators understand and find out the contributing factors in the accidents.

1.7 Implementation of safety management system

At almost the same time as major accidents had been continuously occurring and human factors and SCM were being developed, there was another important change in the international shipping industry. In the wake of the *HERALD OF FREE ENTERPRISE*, *EXXON VALDEZ* and

SCANDINAVIA STAR accidents, the international management code for the safe operation of ship and pollution prevention, called International Safety Management Code (ISM Code), was adopted as Resolution A.741(18) in 1993 (IMO 1993).

The ISM Code became mandatory by entry into force of SOLAS Chapter IX in 1998. Ship management companies are obliged to establish and monitor the safety management systems which satisfy the requirements of the ISM Code. The ISM Code is based on ISO 9001 which is to enhance a product quality by the model of a process-based quality management system: The methodology is known as "Plan-Do-Check-Act" (PDCA) cycle.

The ISM Code was hastily adopted by the international shipping industry after the multiple high-profile accidents detailed above. By the year of 2015, it had experienced five amendments since coming into force.

1.8 Development of comprehensive accident investigation

International efforts had been formed in the way of Resolutions adopted by IMCO and then IMO to reduce the number of marine accidents, especially in the early days of oil pollution at sea since the *TORREY CANYON* casualties. The content and the effects of those resolutions were limited in the scope of approach and methodology of accident investigations.

In order to coordinate the marine accident investigations, Cooperation in Maritime Casualty Investigations was adopted by IMO as Resolution A.637(16) in 1989. However, the resolution had some shortcomings in that it was not mandatory, only the state conducting the investigation had the right to decide if or when to publicize a casualty report, and it emphasized the role of the flag state without giving equal importance to the role of substantially interested states, such as coastal states, of which the definition was also unclear. The resolution did not include unified procedures on safety casualty investigation, nor common approach and methods for the investigations.

Amendments to the code for the investigation of marine casualties and incidents, Resolution A.849(20) and A.884(21) were adopted in 1997 and 1999 respectively. Resolution A.849(20) was the first code on accident investigations lead by IMO. The code was formally adopted on 16 May 2008 in Resolution MSC.255(84), and was made mandatory by adding regulation 6 in SOLAS Chapter XI-I: Special measures to enhance maritime safety.

The code, called Casualty Investigation Code (CIC), aims to achieve:

(1) separation of the safety investigation process from disciplinary or criminal proceedings,

(2) qualified indemnity in disciplinary and criminal proceedings,

(3) confidentiality and anonymity,

(4) rapid, transparent, impartial, objective and accurate reporting,

(5) a simple reporting format which is followed by all countries,

(6) publication and wide dissemination of reports and findings, and

(7) consistent data input to IMO.

The CIC went into effect on 1 January 2010. Furthermore, Resolution A.1075(28) on Guidelines to assist investigators in the implementation of the CIC was adopted in 2014. The guidelines were produced mainly because the safety and protection of the marine environment can be enhanced by timely and accurate reports identifying the circumstances and causes of marine accidents and incidents, and the investigation and proper analysis can lead to greater awareness of casualty causation and result in remedial measures, preventing similar cases in the future.

The entry into force of the CIC urged the establishment of a safety investigation authority in the world.

1.9 The COSTA CONCORDIA disaster

Despite circulation of lessons learned from the past accidents issued by IMO and other organizations, 100 years after the *TITANIC*, the *COSTA CONCORDIA* disaster occurred on 13 January 2012 (Italian Ministry of Infrastructure and Transport 2012).

The *COSTA CONCORDIA*, Italy flagged passenger vessel, had left the port of Civitavecchia, Italy, on 13 January 2012 and was navigating to Savona in the Mediterranean Sea with 3,206 passengers and 1,023 crewmembers on board. The vessel was sailing too close to the coastline in a poorly lit shore area under the master's command who had planned to pass at an unsafe distance during night time and at 15.5 knots. The vessel collided with the Scole rocks at the Giglio island at 21:45. The vessel immediately lost propulsion and was consequently affected by a blackout. The emergency power generator switched on as expected but was not able to supply the utilities to handle the emergency and worked in a discontinuous way. The rudder remained blocked completely starboard and no longer handled. The vessel turned starboard and finally grounded, due to wind and current, at the Giglio island at around 23 m depth and seriously heeled, approximately 15 degrees. After about 40 minutes, the water reached the bulkhead deck in the aft area. The master did not notify the search and rescue (SAR) authorities of his own initiative, and despite the SAR authorities contacting the vessel he informed these authorities about a breach only at 22:26, launching the distress at 22:38. The abandon ship was ordered at 22:54 but it was not preceded by an effective emergency alarm. The first lifeboat was lowered at 22:55. Crewmembers, the master included, abandoned the ship at about 23:20, while one officer remained on the bridge to coordinate the abandon ship. All the saved passengers and crewmembers reached Giglio island since the vessel had grounded just a few meters from the port of Giglio. The first rescue operations were completed at 06:17 the next day, saving 4,194 persons. The rescue operations continued and on 22 March the last victim was found. Environmental operations immediately took place for recovering oils.

The accident investigation report pointed out human factors, namely liveware-software, liveware-peripheral liveware issues, among other contributing factors.

(1) Usage of inappropriate chart, 1/100,000 size scale chart, instead of 1/50,000.

(2) The passage of 0.5 miles off the coast.

(3) No challenge to the master to warn him to accelerate the turn.

(4) Distraction due to presence of extraneous person, hotel director, at the bridge and phone call.

The most important element of making a passage planning is risk assessment to the waters where the master intended to navigate. The master in this case made a planning of a touristic sailing course up to a bathymetric contour line at 10 m depth.

1.10 Similar accidents and insights

When looking at the core of the contributing factors in grounding of the *COSTA CONCORDIA*, a similar case of grounding had already occurred in 1995.

The *MARIA ASUMPTA*, a UK flagged sail-training vessel, departed Swansea, Wales, with 14 people onboard, to the port of Padstow, Cornwall, England, on the morning of 30 May. The master, who was a naval officer and sailed tall ships for more than 20 years, knew the coast and his vessel intimately. He laid a route inside Newland, which was not recommended by the Admiralty, the most commonly used sailing directions. The reason was that he had been accustomed to smaller margins of safety, intended to avoid the effect of a tidal stream off the coast, and wanted his passengers to enjoy the attractive coastal scenery.

As the vessel approached close to the Mouls, a small island, a crewmember suggested tacking out to sea to get clear of the shore. The master decided to sail through the narrow water between Roscarrock and Rumps Point. Influenced by wind and tidal stream, the vessel was closing ashore, and then the master started the engines. Five minutes later, the engines stopped and could not be restarted. Although lookouts had been posted at the bow, they could not spot submerged rocks, the vessel struck rocks at Rumps Point. Three members of the crew were drowned.

Two vessels, the *TITANIC* and the *MARIA ASUMPTA*, were different sizes and types of vessel. On the other hand, both masters, who had knowledge, experience, and were familiar with the waters, apparently underestimated the principle of passage planning that Swift (2000) showed.

"The ships always remain in safe water, sufficiently far off from any danger to minimize the possibility of grounding in the event of a machinery breakdown or navigational error."

1.11 The number of accidents decreased?

Table 1.1 shows the number of marine accidents that occurred, in ocean-going vessels to which ISM Code was applied, in and around Japanese water from 1994 to 2014 (JCG 2015).

Although ISM Code became mandatory for almost all types of vessels of 500 gross tons or more in 2002, the accident ratio in ocean-going vessels remained stable. Considering the trend of accident ratio on the table, current ISM Code has not been very effective in reducing the expected number of accidents.

An intensive research found that the "Plan" and "Do" parts of the PDCA cycle were not afforded by ship management companies that got involved in the accidents in Japan (Fukuoka 2016a).

In history, there have been continuous international efforts to address marine accidents on unified accident investigation, increasing

Table 1.1: Trend of marine accidents in and around Japanese water.

Year	Number of foreign vessels entered at ports in Japan (A)	Number of accidents occurred on foreign vessels (B)	Ratio of B/A (%)
1996	–	298	–
1997	103,796	255	0.24
1998	99,048	248	0.25
1999	103,402	218	0.21
2000	106,640	261	0.24
2001	106,717	261	0.24
2002	105,144	255	0.24
2003	109,120	264	0.24
2004	109,916	277	0.25
2005	108,179	257	0.23
2006	111,172	253	0.22
2007	108,949	260	0.23
2008	102,040	259	0.25
2009	92,721	268	0.28
2010	98,697	239	0.24
2011	96,109	208	0.21
2012	99,740	238	0.28
2013	97,097	229	0.24
2014	95,673	216	0.23

number of establishments of independent marine accident investigation organizations, implementation of safety management system, development of human factors and accidents models that fit to marine domain as contributing factors of accidents have shifted from hardware issues to software and system issues in a socio-technical context. Common approaches, including accident models and accident investigations into organizational factors by safety investigation organizations, have been moving forward.

However, the number of marine accidents has not been dramatically decreasing despite efforts by IMO, organizations such as Marine Accident Investigators' International Forum (MAIIF) and marine accident investigation authorities. Similar accidents have been occurring across the world at any time.

1.12 Conclusions

To reduce the number of marine accidents as well as to establish proactive measures, important element of breakthrough in systematic accident prevention have all been in lined up in the 2000s: Scientific accident investigation, human factors and accident models, and already verified principles on safety and statistics. The remaining issue is the question of how to coordinate and integrate those knowledges.

Chapter 2

Mechanism of Accident Occurrence

2.1 Concept of hazard and accident

Concept of hazards is useful considering an overview of progressive process of the accident and its preventive measures. According to ISO/IEC Guide 51 (1999), hazard is a potential source of harm, and a hazardous situation is a situation where people, property and environment coexist. An accident occurs when, for instance, people make an error in a hazardous situation, or harmful event, resulting in damage to people, property and environment, called harm. Between these situations and events, there could be a layer of defense that helps to avoid an accident (Fig. 2.1).

Apply this concept to the process of an accident which often occurs on a chemical tanker across the world, where crewmembers lost their lives because they entered a cargo tank without detecting atmosphere in the cargo tank filled with toxic gas, or with lack of oxygen. The tank hatch is equipped as a defense between people, property, environment and a hazardous situation. The safety management system (SMS) manual, on entering an enclosed space, exists as a defense between a hazardous situation and a harmful event. The SMS manual prescribes that atmosphere in the cargo tank be measured at different levels of height from the inner bottom plating using a gas/oxygen concentration detector prior to entering the enclosed space. As a defense between harmful event and harm, an emergency response is stipulated for when a crewmember has entered without proper procedures; detecting the atmosphere in the cargo tank and wearing personal protective equipment (PPE). For instance, one

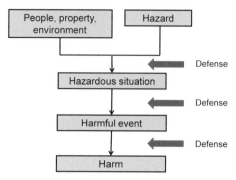

Fig. 2.1: Concept of hazard and accident.

crewmember is to remain on the deck monitoring the colleague working in the cargo tank. When facing an emergency, he is to inform the master and prepare to rescue his colleague with PPE. A rescue team consists of trained crewmembers.

This kind of accident happens repeatedly, everywhere in the world, every year. To figure out the cause of occurrence, it is necessary to think about the effect of a layer of defense.

2.2 Effect of a layer of defense

Figure 2.2 shows a priority ranking in risk reduction: (1) inherently safe design, (2) protective devices, (3) information for safety, and (4) administrative control and PPE. Risk is reduced gradually by taking these steps, but it remains even if the last step is reached (ISO/IEC 1999, OHSAS Project Group 2007).

Steps (1) to (3) are for product designers and manufacturers, step (4) is for the user. Users in the maritime industry include ship management companies, ship owners, ship operators and crewmembers. When a ship management company considers risk reduction methods, first, inherent safety design is considered; this includes either elimination, which is to remove people from immediate contact with a hazard, or substitution, which is to replace the hazard with a safe alternative. Next, engineering control is considered, which involves developing new equipment or modifying existing equipment in order to protect workers. These works can require the involvement of crewmembers or equipment designers and a dockyard. Then we have signage, warning and administrative control. Administrative control includes establishment and maintenance of safe

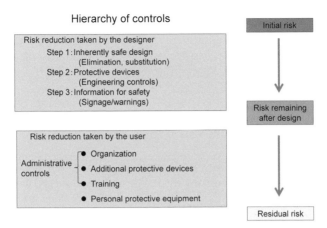

Fig. 2.2: Priority ranking in risk reduction methods.

work procedures as well as education and training of workers in hazard control. Finally, PPE must be worn.

One of the points to reduce the number of accidents and incidents is for all maritime community to understand this hierarchy of control and comprehend the signage, warning, administrative control such as SMS manual, and donning PPE are least effective among the risk reduction methods.

Apply the case of chemical tanker accident to this hierarchy of control. The tank hatch is not regarded as an inherently safe design but as a protective device. During unloading or tank cleaning, many crewmembers of the vessels or workers who were employed by external companies have been losing their lives upon entering into the cargo tank filled with toxic gas or low oxygen concentration with inappropriate PPEs. In the past, even colleagues attempting to rescue those who had become unconscious on the inner bottom plating were involved in an accident and suffered a second disaster because of failing to measure the atmosphere and wear appropriate PPEs.

International efforts to warn the recurrence of this kind of accident have been made by IMO, MAIIF, marine accident investigation authorities and other organizations, and IMO adopted the procedures on entering enclosed spaces in 2011 (IMO 2011), and posted lessons learned drawn from the accident investigations from member states on its website. MAIIF circulated the leaflets that depict the danger of hiding out in enclosed spaces. However, as shown in Fig. 2.2, these measures are classified as the least effective in the hierarchy of controls.

It can be said that an inherently safe design would solve this issue: If possible, such a system should be equipped onboard so that the tank hatch can be opened only when the atmosphere measured in the tank is within the acceptable level. Without it, the maritime industry will have to continue relying on the least effective way of risk reduction.

2.3 Situation of defenses in depth and accidents

With advances in technology and complexity in recent socio-technical society, the situation of defenses in depth and human factors is deeply related to the contributing factors in most of the accidents and incidents. Commonly used principles on safety, shown on Figs. 2.1 and 2.2, cannot provide any insights into the situation of defenses in depth, human factors and their relationship with accidents and incidents.

The SCM developed by Reason (1997) can explain the situation of defenses in depth. The SCM is used as background theory in the marine accident investigation field, and it displays that there are defenses in depth between hazards and losses or an accident. Although it is ideal that defenses in depth are without defects, there are problems, such as defects occurring during the product design process and operating process, because designers, shipbuilders, safety managers, operators or crewmembers cannot foresee all possible accident scenarios. This defect is called "a hole" on a defense, which is moving around and changing its size, opening and shutting. The holes arise due to active failures and latent conditions. The active failure is an error made by an operator in a hazardous situation that leads to an accident, and latent conditions are conditions that lie dormant until the accident occurs, poor design, inappropriate supervision, undetected flaws during the manufacturing process, failure of maintenance, and unfunctional procedures, for example. An accident occurs when holes on defenses in depth line up, an accident trajectory goes through the holes, and then causes people, property, and environment to have direct contact with hazards and harm. The SCM, like other accident models, is not flawless; this is discussed further in Chapter 8.

Apply again the case of chemical tanker accident during enclosed space entry to SCM. It is apparent that the procedure on enclosed space entry was not followed onboard or it did not exist from the beginning. Those are latent conditions related to the organization: Ship management company. The crewmember opened the tank hatch without detecting the atmosphere in the cargo tank or without wearing an appropriate PPE.

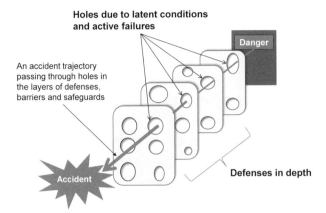

Fig. 2.3: The Swiss cheese model (Reason 1997).

This is an active failure related to the operator. The SCM shows that the holes opened in defenses, the accident trajectory went through the holes and an accident occurred.

The active failures and latent conditions are related to actions or conditions in which humans are involved. The SCM clarifies that holes arise in defenses but does not explain the implication of human factors and contributing factors in detail.

2.4 Human factors and accidents

The SHEL model is widely used both by marine and by aviation accident investigation authorities as a human factor tool, not only for collecting evidence on the accident site, but also for analyzing the contributing factors of the accidents (Fig. 2.4).

The SHEL model consists of five components: Central liveware, hardware, software, environment and peripheral liveware. Central liveware deals with the condition of operators by themselves at the time of an accident. Liveware-hardware interface relates to the man-machine system. Liveware-software interface refers to non-physical aspects of the system, such as SMS manuals and procedures. Liveware-environment interface deals with environment and human requirement. Central liveware-peripheral liveware interface is concerned with an operator and other people. In this model, central liveware is the hub to which the remaining components, i.e., software, hardware, environment

17

and peripheral liveware, must be matched. A mismatch can be a source of human error that contributes to the occurrence of an accident. Table 2.1 shows the overview of each element of SHEL, which is adapted from Hawkins and ICAO to fit the marine field (Fukuoka 2016b).

Fig. 2.4: The SHEL model (Hawkins 1987).

Table 2.1: Overview of the SHEL model in marine field.

Element of SHEL	Definition
S	Relationship between the individual and supporting systems, or non-physical part of the system found in the workplace. Includes regulations, manuals, checklist layout, charts, maps, publications, standard operating procedures, computer software design, etc.
H	Relationship between the human and the machine. Includes design and condition of workstations, displays, controls, seats, and all other physical parts of a ship or system, etc. Also includes any problem with experimental devices and equipment/engine of ship.
E	Relationship between the individual and the environments. The environment includes noise, temperature, light, vibration, motion, air quality, and atmosphere in which people work. Also includes weather, visibility, sea conditions, traffic density, etc.
L	Includes physical factors (physical limitations of the individual, visual/auditory other sensory limitations, light adaptation, glasses and contact lenses), physiological factors (health condition (disease), prescriptions or drug, fatigue (amount and quality of sleep, work-rest schedules, Circadian rhythm)), psychological factors (interruption, distraction, assumption, stress, anxiety), psychosocial factors (death of spouse, divorce, financial problems, etc.), experience, knowledge and training.
L (Peripheral)	Relationship between people. Consider language barriers, misleading communication, quality of communication equipment, teamwork, cultural and language differences, adequacy of supervision, task assignment, workload distribution, information dissemination, etc. Involves safety management and safety culture.

The SHEL model does not include technical failures which are not linked with human factors and are regarded as external factors. In a real world, accidents related to technical failure have been occurring, such as a case where the center of gravity became higher due to alteration of the equipment of the ship, and the ship's stability deteriorated, resulting in a capsizing while navigating in adverse sea and weather conditions. In Table 2.1, technical failures are added to hardware to make up for the issue.

2.5 Conclusions

It is noted that simply combining the SHEL model and the SCM will not be enough to find out where a hole arises in defenses in depth and how the relationship between a hole and latent conditions are closely related. In order to clarify these issues, it is necessary to analyze the progress of development of the accident models in all fields and the actual marine accidents in detail, comparing them with the SHEL model and the SCM.

Chapter 3

Accident Model

3.1 Background surrounding accidents

Accidents occur in every industry, incurring great loss of life and property. For instance, in maritime transport in Japan, more than 2,000 vessels were involved in marine accidents, such as collisions, groundings, fire, etc., and the lives of about 80 to 140 people have been lost in territorial waters and surrounding waters each year (JCG 2015).

Marine accidents not only incur losses of life and property, but also cause severe environmental pollution and tarnish the reputation of the company related to the accidents, such as in case of the grounding of the *TORREY CANYON* (about 119,000 tons of oil spill) in the Dover Strait in 1967, and the grounding of the *EXXON VALDEZ* off the coast of Alaska, USA in 1989 (40,882 tons of crude oil spill).

In the airline industry, according to ICAO accidents statistics on scheduled commercial flights of airplanes above 5.7 tons, the number of accidents has fluctuated between 75 and 140 from 2001 to 2018. In the electric power industry, a major blackout occurred in the Northeastern and the Midwest in the USA and in Ontario of Canada in August 2003, resulting in a total of 50 million people affected and the urban function failed. In the nuclear power industry, the Fukushima Daiichi nuclear power plant disaster occurred in the wake of the major earthquake and subsequent tsunamis that struck northeastern Japan in March 2011. In addition to loss of life and property, extensive communities inhabiting the land around the plant were forced to evacuate from their homes for a long time.

Once an accident like these occurs, many lives, property, environment, reputation of the company and urban function are lost and serious damage is caused; therefore, prevention of accidents is an important issue in all sectors of industries.

To prevent accidents, it is necessary to conduct a scientific accident investigation based on evidence and to analyze contributing factors using appropriate accident models. In this chapter, with a prerequisite that scientific accident investigation is being conducted, focusing on the SCM which is widely used in marine domain, both historical background of the accident model and the field of application of the accident model are described.

3.2 Historical background of accident model

The accident model is important as a tool to understand phenomena, such as accidents or potentially dangerous system behavior, and as a tool to share contributing factors when discussing accidents with other people (Leveson 2011).

According to Hollnagel (2004), the accident model is a routine way of thinking about how accidents occur, and is classified as a sequential accident model, an epidemiological accident model, and a systemic accident model. The principle of each accident model, the goal of the analysis, the characteristics of the analysis, the process of the accident analysis, and examples of the representative accident model are shown in Table 3.1 (Hollnagel 2004, Hollnagel and Speziali 2008).

3.2.1 Sequential accident model

The sequential accident model is represented by the domino theory advocated by Heinrich, and it is also referred to as the domino model. As the five factors of event chain, social environment, fault of person, unsafe act, accidents, and injury, are depicted in a line, if the occurrence of the event in the middle in the sequence is blocked, it is said that the accident can be prevented (Dekker 2006). Fault tree analysis (FTA), Failure mode and effect analysis (FMEA), and Why-why analysis are methods to analyze accidents from the same perspective as the domino theory.

Table 3.1: Classification of accident model.

	Sequential accident model	Epidemiological accident model	Systemic accident model
Search principle	Specific causes and well-defined cause-effect links	Carriers, barriers and latent conditions	Tight coupling and complex interactions
Analysis goals	Eliminate or contain causes	Make defenses and barriers stronger	Monitor and control performance variability
Features	Describe accidents as a causal series. Suitable for linear systems wherein the cause-effect links are relatively simple. Adequate for socio-technical systems until the middle of the 20th century. Able to prevent accidents if a root cause is corrected	Analogize an accident with the spreading of a disease. Describe accidents as a causal net. Follow the principles of the sequential model; cause-effect links. Performance deviation is bad, and caused by error-promoting conditions. Able to prevent accidents if error-promoting conditions and barriers are corrected	Consider accidents as normal. Eliminate cause-effect links. Performance variability is necessary for a user to learn and for a system to develop. Monitoring of performance variability is needed to distinguish between what is useful and what is harmful
Process of accident analysis	Trace the development of events back in time, starting from the accident as it occurred	Trace the development of events back in time, starting from the accident as it occurred	Trace the development of events back in time, starting from the accident as it occurred
Examples	Domino model (Heinrich), Fault tree analysis, Event tree analysis, Fishbone analysis, Why-why analysis, Accident Evolution and barrier analysis, Root cause analysis, Human error in European air traffic management, Failure mode and effect analysis, Sequentially timed events plotting, etc.	Swiss cheese model (Reason), SHEL and Reason hybrid model, Human factors analysis and classification system (Wiegmann and Shappell), ATSB model	Functional resonance accident model (Hollnagel), Systems-theoretic accident model and processes (Leveson)

In the 1970s and 1980s, a number of such major accidents involving socio-technical systems occurred, such as the Flizborough chemical plant accident in the UK (occurred in 1974), the Three Mile Island nuclear power plant accident in the USA (1979), the Japan Airlines crash in Mt. Osutaka in Japan (1985), the Space Shuttle Challenger explosion accident in the USA (1986) and the *HERALD OF FREE ENTERPRISE* accident in Belgium (1987). As a result of investigating and analyzing these accidents in detail, it became clear that human error could not be explained without considering the circumstances surrounding the operators who caused the error, such as local workplace factors and organizational factors, hence, accident models of a different type to the sequential accident model were required (Reason et al. 2006).

3.2.2 *Epidemiological accident model*

The SCM as a representative of epidemiological accident model was developed in 1990. The SCM makes the accident occurrence process similar to the relationship between human health and pathogens, successfully explaining the background to the organizational accident and the background theory of the accident, being used as a method to analyze and prevent organizational accidents occurred in industrial fields, such as ships, aircraft, railway, road transportation, and medicine (Anca 2007). The SCM categorizes unsafe acts from a psychological point of view and incorporates the concept of latent conditions which weaken the defensive layers and trigger active failures at front end due to human factors, safety culture, and so forth (Reason 1990, 1997).

In addition to the SCM proposed by Reason, there are several epidemiological accident models. The SHEL and Reason hybrid model that combines the SCM with the SHEL model, the Human factors analysis and classification system (HFACS) that developed from the SCM, and the ATSB model that was developed by the Australian Transport Safety Bureau (ATSB) and used for analysis of vessels, aircraft and railway accidents (ATSB 2008).

According to the theory of the SCM, an accident will not happen if a hole of the arbitrary defensive layer is closed so that the accident trajectory will not be able to go through the whole defenses in depth. By identifying the location of hole and clarifying the relationship with latent conditions, it is possible to close the hole, that is, to prevent an accident.

3.2.3 Systemic accident model

In the systemic accident model, it is assumed that an accident occurs when the components of the sociotechnical system fluctuate, excluding the concept of accident causes from the model. It is represented by the functional resonance analysis method (FRAM), and Systems-theoretic accident model and processes (STAMP). The systemic accident model was developed on the grounds that the epidemiological accident model cannot explain the accident that occurs when a normal person performs the same work as usual. It focuses on all the elements that make up the sociotechnical system by eliminating the link between the cause and the result of accident occurrence, and emphasizes that accidents occur when problems occur in the interaction of each element.

Dekker (2006) points out the accident investigation result of the Space Shuttle Columbia accident that occurred in the USA on 1 February 2003, regarding the necessity of a systemic accident model. He states that, although physical or mechanical factors can be explained by the sequential accident model and the epidemiological accident model, organizational background cannot be explained. That is, it is argued that the epidemiological accident model cannot explain why engineers in the organization came to think of actions or system performance that had been deviated from usual as ordinary or acceptable.

Dekker states that the systemic accident model has the following two features. The first, regarding safety, is an emergent property that indicates that the constituent elements of the system are well matched with other constituents and the whole is functioning. The other is that, although the mutual relationship between the constituent elements is controlled by the procedure, such as procedure manual and design requirement, the sociotechnical system will fluctuate so that the set control does not conform to the latest situation. The systemic accident model considers human error to be necessary for system development and purpose, not to be the main analysis subject, and the problems caused by the human factor and safety culture are analyzed in the framework of emergent property and control, as shown below in the STAMP analysis method. Therefore, the systemic accident model, unlike the sequential accident model and the epidemiological accident model, determines that accident occurrence is a byproduct of normal function and puts emphasis on constantly monitoring the constituent elements of the sociotechnical system and evaluating whether the control is functioning.

3.3 Problems on each accident model

3.3.1 *Sequential accident model*

The sequential accident model is characterized by a sequence of events that occur in a specific order. It can be applied only to accidents or industrial fields where the accident factors of the system can be physically, clearly tracked down and not to accidents that occurred in a sociotechnical system in which the interaction between the system components is complicated and difficult or impossible to track down.

Leveson (2011) states that, at present, there is a limit to risk assessment in the sociotechnical system as follows: (1) Although technological innovation has drastically changed the environment surrounding safety engineering, and despite the fact that interaction among system components has become complicated, engineers use FTA and FMEA, which are applied to analysis of linear systems, and the effectiveness of safety engineering is decreasing; (2) Analysis Probabilistic risk assessment uses a linear sequential accident model when quantifying risks.

3.3.2 *Epidemiological accident model*

The problems of the epidemiological accident model are described in relation to the HFACS, the SHEL and Reason hybrid model, the ATSB model and characteristics of SCM. In this chapter, the term "SCM" refers to a model including the "Early SCM", "Middle SCM", "Late 1 SCM" and "Late 2 SCM" described below. When referring to a specific period of model that was released by Reason, "Early", "Middle", "Late" are attached before SCM. The SCM described in Fig. 2.3 in Chapter 2 is the Late 1 SCM, when referring to Late 1 SCM in this chapter, revert to Fig. 2.3.

HFACS

HFACS was developed by Wiegmann and Shappell in the process of analyzing accident investigation reports on a number of military aircraft and civil aircraft in the aviation field in order to identify causal factors of the accident. It is applied to fields such as ships and railways as well as aviation, and it is used to decide accident statistics and measures in order to prevent recurrence of accidents. HFACS adapted the Early SCM that was released in 1990, as shown in Fig. 3.1, and classifies the

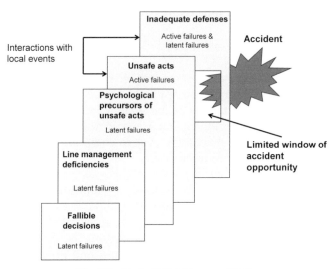

Fig. 3.1: Early SCM (Reason 1990).

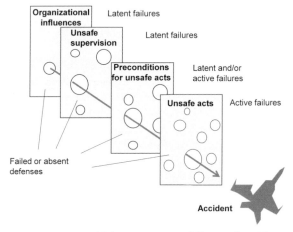

Fig. 3.2: HFACS (Wiegmann and Shappell 2003).

causes of accidents into organizational influences, unsafe supervision, preconditions for unsafe acts and unsafe acts (human error and violation), as shown in Fig. 3.2.

According to HFACS (Wiegmann and Shappell 2003), organizational influences include resource management, organizational climate, such as organizational structure, policies and culture, and organizational process, which consists of operations, procedures and oversight. Unsafe supervision shows inadequate supervision, planned inadequate operations, failure to correct a known problem and supervisory violations. Preconditions of unsafe acts include adverse mental states, adverse physiological states, physical or mental limitations, crew resource management, personal readiness, physical environment, and technological environment. Unsafe acts are based on GEMS framework as skilled-based errors, decision errors, perceptual errors, routine violations and exceptional violations.

Although HFACS uses causal factors conforming to aircraft accidents and many holes are drawn, it is not clear about the reason for the existence of a hole, the characteristics of the hole, such as movement of the hole, the relationship between the hole and the causal factors. Wiegmann and Shappell (2003) acknowledge that the SCM has limitations when it is used for actual aircraft accident analysis, claiming that it is necessary to clarify the hole in order for the model to be used systematically. Reason et al. (2006) point out that Early SCM cannot be applied to analysis of complicated organizational accident because progress in the occurrence of an organizational accident is drawn linearly in the order of fallible decisions, line management deficiencies, psychological precursors of unsafe acts, unsafe acts, and inadequate defenses.

SHEL and Reason hybrid model

As shown in Fig. 3.3, the SHEL and Reason hybrid model is an accident model mainly combining the SHEL model and the Early SCM model, and IMO adopted it as a guideline for a tool of marine accident investigation into human factors until 2014. The process of investigating human factors consists of six steps: (1) collect occurrence data, (2) determine occurrence sequence, (3) identify unsafe acts/decisions and unsafe conditions, (4) identify the error type or violation, (5) identify underlying factors, and (6) identify potential safety problems and develop safety actions.

During steps (1) and (2), the SHEL model and Early SCM are used to comprehensively gather evidence related to the accidents and then to establish occurrence events and circumstances around five production elements: Decision makers, line management, preconditions, productive activities and defense. Step (3) indicates that causal factors are selected among evidence obtained after accomplishing steps (1) and (2), and that causal factors are defined as unsafe acts/decisions and unsafe conditions. During step (3), SHEL and Reason hybrid model are useful for conducting iterative assessment. Step (4) shows that the error type or violation is classified according to Generic Error Modeling System (GEMS)

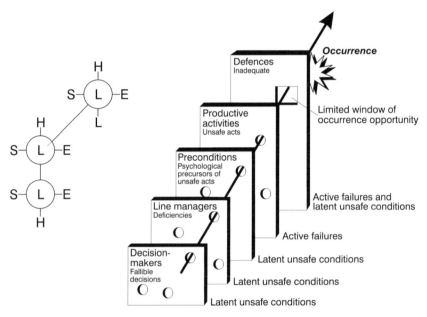

Fig. 3.3: SHEL and Reason hybrid model (IMO 2000c).

framework. And then underlying factors that contributed to the error or violation are identified among evidence collected during steps (1) and (2). Finally, measures are taken to correct safety issues.

ATSB model

ATSB model was developed based on Middle SCM that was released in 1995 (Fig. 3.4). It consists of such five elements as organizational influences, risk controls, local conditions, individual actions and occurrence event, and indicates that an accident occurs when an operator's unsafe act passes through the last defensive layer (Fig. 3.5).

Organizational influences include risk control implemented by the organization, rules and regulations established by the flag state and international standards adopted by IMO and classification societies and so forth that affect the organization. Risk control involves education and training, procedures, BRM, communication, appropriate allocation of the crew, backup system of equipment and alarm devices. The difference between the organizational influences and the risk control is that the former is the risk control that the organization conducts, and the latter is the one

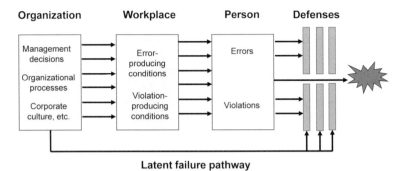

Fig. 3.4: Middle SCM (Reason 2008).

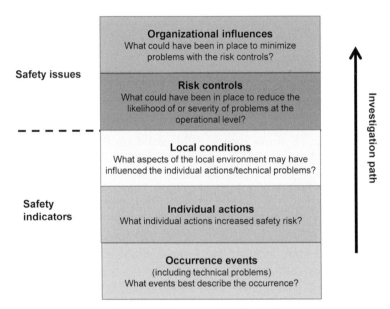

Fig. 3.5: ATSB model (ATSB 2008).

at the local workplace where the operator is active. A local condition is a situation existing immediately before occurrence of individual behavior or technical failure, and it is a situation that affects individual behavior or technical failure. Specifically, the local condition indicates the health condition of the operator, fatigue, stress, drinking and other human factors, operator's experience, knowledge, skill, excessive workload, time pressures affecting the operator, work environments such as lighting, noise, humidity, vibration, atmosphere in enclosed spaces, visibility, wind force, wave height and sea and weather conditions. Individual actions are actions which are performed by operators and level of exposure to risks. The occurrence events indicate what happened to the ship, and the individual actions and occurrence events are closely tied to each other.

Even though the ATSB model, like the SHEL and Reason hybrid models, is based on the SCM, the features of the hole, such as the location of hole and relationship between the hole and latent conditions, are not shown. The ATSB model is based on Middle SCM, which cannot be applied to the analysis of complicated organizational accidents the same way as Early SCM, on the other hand, Underwood and Waterson (2013) state that the ATSB model can be categorized as the systemic accident model because it contains a system thinking approach, which is a requirement to be classified as the systemic accident model described later. The IMO (2014) adopted an accident model with some modifications to the ATSB model at IMO model course 3.11, used for education and training of marine accident investigation.

Late 1 SCM

Figures 2.3 and 3.6 show Late 1 SCM and Late 2 SCM, respectively, published by Reason in 1997. Reason used Late 2 SCM mainly to explain the progress of organizational accidents and the process of accident investigation. In his book, published in 2008, Late 2 SCM is not quoted in the series of development of SCM; therefore, Late 2 SCM is not mentioned in this chapter, but in Chapter 5.

According to Reason, Late 1 SCM was developed by making two major alterations to Early SCM and Middle SCM. For one, in the previous models, each defensive layer was named as fallible decisions, line management deficiencies, psychological precursors of unsafe acts, unsafe acts and inadequate defenses, but in Late 1 SCM, these names are not used, instead, being expressed as defenses in depth as a whole. Also, latent failure or latent unsafe conditions is rephrased as latent conditions.

Reason explains the alteration of the term that unsafe or failure is the word used as the consequence of human acts and is not an appropriate expression. It is considered that the term alteration is related to the fact that failure or unsafe is confirmed after the consequence of event and that, while the operator is doing the work, they do not do it while knowing their actions to be unsafe or failed. Reason explained that the hole opens

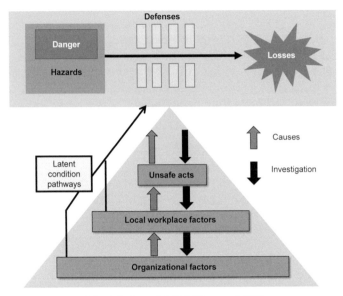

Fig. 3.6: Late 2 SCM (Reason 1997).

due to an active failure or latent conditions, and the latent conditions include poor design, inadequate supervision, undetected defects in the manufacturing process, maintenance failure and unfunctional procedures.

Without assigning names to each defensive layer in Late 1 SCM, the linear process of accident occurrence, which was regarded as a restriction when applying Early and Middle SCM to accident analysis and preventive measures, will be eliminated, and is considered to be applicable to accidents occurred in sociotechnical systems. In Late 1 SCM, similar to Early SCM and Middle SCM, the location of the hole and the relationship between the hole and latent conditions are not shown in detail.

As SCM has changed between 1990 and 1997, it clarifies the existence of local workplace factors and organizational factors behind human acts on organizational accidents and is established as a useful model for thinking about analysis and prevention measures. SCM has some unsolved issues: What a hole is composed of, why the hole opens, how the hole moves in defensive layer, why the hole has to line up in order for an accident to occur (Wiegmann and Shappell 2003, Dekker 2006, Reason et al. 2006). The same is true for HFACS, SHEL and Reason hybrid models and the ATSB model developed from the same model.

Reason et al. (2006) state that SCM has been established as a standard scientific method and is currently being used for accident analysis in many fields. They acknowledge that no significant experiments, such as denying SCM, had been carried out and that if it could identify the latent conditions that lay dormant in the system in advance, measures could be taken to prevent accidents from happening.

3.3.3 Systemic accident model

STAMP (Leveson 2011), which is a representative of the systemic accident model, is an accident model that adds system thinking and human factor engineering, developed from the system safety of aerospace systems, such as military aircraft and ballistic missiles.

The outline of the analysis method is as follows: (1) Identify loss-related systems and hazards, (2) Clarify the purpose and constraints of the system components related to the hazard specified in (1), (3) Identify the most immediate event that led to the loss and analyze the loss that occurred at the level of the physical system. That is, clarify contributing factors, such as ineffective operation management, physical failure, interaction that caused functional failure, defect of communication and adjustment, unprocessed failure, etc., (4) Clarify inadequate management that contributed to the loss in (3) on the high level of safety management system that oversees (3), (5) Investigate the condition of adjustment related to loss and communication on all components of the system and clarify the contents of the system and the safety management structure which changed over time, (6) For all system components, clarify safety requirements and constraints, situations in which decisions were made, insufficient control actions, and mental model defects. Describe these in the diagram composed of each constituent element and connect the management system channel and the communication channel networked between the constituent elements with arrows, (7) Take measures to prevent recurrence in order to correct defects and contradictions that occurred in the system clarified by the analysis of (6).

A system thinking approach is important when analyzing accidents occurring in sociotechnical systems. The system thinking approach is made up of the following three requirements: (1) The hierarchical structure of the system and its boundary are indicated, (2) All components that make up the system and the interrelationship between these components should be taken into account, (3) It shows the goals, resources, and circumstances of economic activities affecting human behavior.

The systemic accident model satisfies these requirements, but epidemiological accident models, such as SCM, do not meet the requirements of the system thinking approach, so they are distinguished from systemic accident models (Underwoods and Waterson 2013). Underwood and Waterson evaluate the systemic accident model as follows. Since the systemic accident model is required to investigate the general situation of the sociotechnical system in addition to the situation at the time of the accident, the systemic accident model is used exclusively by researchers, and the use is not spread to accident investigation authorities and company safety officers who have temporal constraints on the preparation of accident investigation reports.

3.4 Characteristics of each industry

Perrow (1984) studied and illustrated the characteristics of each industry field in Fig. 3.7. The vertical axis represents the degree of closeness of the coupling between the system and the components, and the horizontal axis represents whether the interaction of the system or the process is linear or complicated. Perrow defines a collection of units, parts and subsystems as constituent elements, and those constituted by a plurality of constituent elements as a system.

Tightly coupled system is synonymous with coupling of engineering terms, there is no slackness, the design of the whole process is the only way to achieve the objective, the sequence of processes is invariant and time-dependent. In the loosely coupled system, on the other hand, there are many ways to achieve the objective, substitutes can be used for units and parts that make up the system, the sequence can be changed, the process delay is flexible. For example, nuclear power plants belonging to the former tightly coupled system cannot change fuel from nuclear to other substances, but in thermal power plants in a loosely coupled system they can substitute petroleum for coal or coal for petroleum. In a loosely coupled system, for instance, even if there is a problem in the manufacture of the door or the seat in the assembly line in the manufacturing factory of the automobile, etc., the finished product can be produced by replacing the parts or the like which conform to the standard later. At nuclear power plants, that is, in a tightly coupled system, the finished product cannot be replaced later.

A linear interaction is predictable and evident in this interaction, regardless of the units, the number of parts, so that even if a malfunction occurs in a unit or the like, it is possible for a person to confirm and

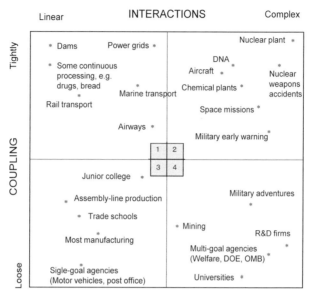

Fig. 3.7: Classification of each industry (Perrow 1984).

predict what is occurring in the upstream and downstream processes of the failure location. In a linear system, manufacturing plants, for example, where a worker misunderstands the instruction and performs work, the work supervisor can notice the mistake because error and interaction of the worker will be visually recognized.

A complex interaction is such that a component intentionally or unintentionally has a plurality of functions at the design stage, so that when a malfunction occurs, it is impossible to visually recognize what is occurring, and since there is no way to guess the situation, it is impossible to grasp the whole situation immediately after the occurrence of the failure. In such a complicated system, not only the unexpected interdependence between components is revealed due to malfunction, but also because the roles and knowledge of persons working here are specialized, it is impossible for people to predict, notice, and analyze this interdependence.

3.5 Accident model applicable to each industry

Hollnagel and Speziali (2008) changed the interaction of the horizontal axis in Fig. 3.7 to the traceability of tracking whether it is easy to describe the system, and deleted continuous processing, multi-goal agencies, nuclear weapons accident, and DNA, and instead added financial markets to the second quadrant. They moved military early warning to the first quadrant, airport and military adventures to the second quadrant, and mining industry to the third quadrant in order to illustrate the applicable range of each accident model.

Tractable, meaning that the functional principle of the system or process is known and simple, represents that the description and the system are consistent. Intractable means that some or all of the functions of the system or the process is unknown, the description is dense, and the system may be changed before the description is complete.

According to the classification of Hollnagel and Speziali, the sequential accident model can be applied to the industries classified in the third quadrant, which include manufacturing industry by assembly line, the majority of manufacturing industry, trade schools, junior colleges, post offices and mining industry. The sequential accident model is applied only to industries in which the cause and effect is clear, that is, in an industry in which the interaction between the constituent elements is linear. It cannot be applicable to the industries where the interaction is complicated because the factors leading to the accident cannot be traced.

The epidemiological accident model can be applied to the industries in the first quadrant, which include dams, power grids, marine transportation, railway transportation and military early warning. Late 1 SCM, which is the epidemiological accident model, can be applied to industries where the interaction is complicated because linear elements are excluded; however, it has limitation in application to the socio-technical system in general because there are cause-effect links.

The systemic accident model can be applied to industries in the second quadrant, which include nuclear power plants, chemical plants, space missions, financial markets, airlines and airports.

Hollnagel and Speziali analyze the reason why the accident model changes over time as follows. The working environment of people changed from production using simplified machines to production activities with complex systems making full use of computer technologies from the industrial revolution to the modern socio-technical system, the form changed from a simple thing with identifiable cause-effect links and with few causal factors to a complicated one whose cause-effect links are not clear due to the involvement of many causal factors. As a result, the accident model was also developed in order to be able to adapt to these changes and shifted from the third quadrant to the first quadrant and the second quadrant, as shown in Fig. 3.8.

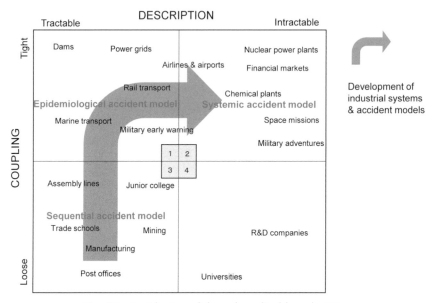

Fig. 3.8: Accident models and applicable industries.

3.6 Conclusions

Considering the historical background of the accident model and the characteristics of the industry as one requirement to select an accident model suitable for each industry, it is necessary to clarify the characteristics of the industry to which the company belongs and select an applicable accident model. Although the accident model applicable to marine transport is the epidemiological accident model as described above, the engine malfunction of the ship is similar to the accident occurred in the manufacturing industry in the third quadrant, hence, the sequential accident model can be used. Collisions, contact, grounding, occupational casualties, fire, explosions, sinking and capsizing have been occurring in a complicated socio-technical system, and it is appropriate to utilize Late 1 SCM suitable for the analysis of industrial accidents in the first quadrant. Accident investigators and analysts in each industry should acquire analytical methods for each accident model and consider it important to use the accident models properly according to the type of accident in addition to the characteristics of the industry to which the company belongs (Fukuoka 2017).

Chapter 4

Contributing Factors of Accident Occurrence

4.1 Introduction

Marine accident investigators utilize the SHEL model for evidence collections and analysis of causal factors in accordance with IMO resolution. The SHEL model is composed of central liveware, hardware, software, environment and peripheral liveware. This chapter looks into subclassification of each element of the SHEL with actual marine accidents and statistics that reveal a variety of contributing factors.

4.2 Central liveware

Central liveware is the most valuable and the most flexible component in the system. It is divided into four categories: Physical factors, physiological factors, psychological factors and psychosocial factors.

4.2.1 Physical factors

Physical factors deal with the physical limitations of the individual. Physical limitations affect the ability to see, act, move, reach and grab. Visual, auditory, and other sensory limitations influence an individual's performance. Studies by Baltes and Lindenberger (1997) show that the ability of eyesight, audibility, perceptual speed and memory decrease as the individual gets older, especially around 60 years old and above.

Human information processing

There are many models and theories about process of human information, but overall five processes are taken as follows: (1) Sensing, (2) Perception, (3) Decision-making, (4) Action, and (5) Feedback (Hawkins 1987, Endsley 2000, Yokomizo and Komatsubara 2013).

Sensing

Information is sensed by sense organs which are not designed for detecting all information that may be important. Sensory system functions with a narrow range and needs more than the threshold of stimulus. Some seafarers experience obstructed vision by glare at sea, and others have collisions when unaware of an approaching vessel due to light adaptation.

Light adaptation, from daylight vision to full night vision, occasionally acts as a contributing factor in cases of collisions. There was a case of collision when the captain of fishing boat *KOSEI-MARU* was not able to detect an approaching merchant vessel *MEDEA* just after finishing preparation for fishing gear at fore deck with strong light shed near the fishing ground. Dark adaptation was incomplete before he started slow-ahead. It takes about 30 minutes for humans to fully adapt to the night vision from the daylight (JTSB 2012a).

Perception

Sensory memory or immediate memory stores information in a period of 0.5 to 1.0 seconds. This means that information is fleeting, short lived and unstable. A piece of information given selected attention goes to the central processing unit where short-term memory (working memory) and long-term memory interact. Humans' ability of attention is limited; short-term memory can store information of 9 to 5 chunks. Information stored in short-term memory becomes long-term memory only if it repeated many times. Then information from short-term memory and long-term memory is compared and analyzed, resulting in comprehension of the nature and meaning of the message received. The structure or organization arrangement of elements in the information influences the way humans perceive things that differentiate between what is seen as the figure or object and what is seen as the background. In other words, comprehension is based on the context in which the message is placed.

Therefore, even if stimulus is the same, the perception of the same stimulus by another person is different. When information is insufficient, humans may unconsciously fill in the missing information themselves.

Decision-making

After comprehension of the nature and meaning of the message, a decision is made which is in one single channel and limited capacity. The decision-making channel is being time-shared between different inputs although humans have a vast capacity of sensing information. The decision-making process is intervened by many factors, such as training, experience, emotional or commercial consideration, fatigue, medication, motivation, physical or psychological disorder, etc. As a result, a bottleneck in this process builds up, and false hypothesis or mistaken assumption are an important contributor to human error.

Analysis of information processing by Endsley model

The information processing can be made clearer by using Endsley's situation awareness model. The model depicts situation awareness in the dynamic decision-making process.

Situation awareness in the model consists of perception, comprehension and projection. Comprehension includes integration of multiple pieces of information from the central liveware itself and the rest of the components of the SHEL model and a determination that relates to the operator's goal. Projection is to expect future events and dynamics from current events and dynamics. In situation awareness, space and time that an operator can allow to use are important elements. Then decision, performance of actions and feedback follow as process. Situation awareness, decision and performance of action are influenced by the operator's goals and objectives, memory, experiences and training, stress and workload, etc. The process iterates; decisions are formed by situation awareness and vice versa (Endsley 2000).

4.2.2 *Physiological factors*

Physiological factors deal with the individual as a complex organism encompassing a large array of systems. It includes health, smoking, alcohol, drugs, nutrition, stress and fatigue. Fatigue has been a major contributing factor of accidents in central liveware.

Fatigue

Fatigue is often confused with sleepiness or tiredness. While these might be symptoms of fatigue, it is possible to be dangerously fatigued without feeling sleepy or tired. Fatigue is defined as a reduction in physical and/or mental capability as the result of physical, mental or emotional exertion which may impair nearly all physical abilities including: Strength, speed, reaction time, coordination, decision making and balance (IMO 2001). In aviation, fatigue contributes to 4–7 percent of civil aviation accidents and incidents, and nearly 95 percent of fatigue results from inadequate amounts of sleep (Caldwell and Caldwell 2003).

Factors of fatigue

Fatigue is associated with slower brain activity, changes in eye movements, pupillary responses and other physiological factors. Although no biological markers for fatigue have been discovered, alertness is related to fatigue. The level of fatigue is calculated by alertness, which is measured by electroencephalograph (EEG), electrooculography (EOG) and electromyograph (EMG). There are main three factors that affect alertness: (1) Homeostatic sleep pressure (homeostatic mechanism); (2) Circadian mechanism; and (3) Sleep inertia (Caldwell and Caldwell 2003).

Homeostatic sleep pressure correlates the amount of time awake since the last sleep period. Although homeostatic drive for sleep is low after awakening, it becomes noticeable when continuous wakefulness extends beyond 16 hours a day.

Circadian mechanism is a daily cycle of alertness and sleep and works as an internal body clock, regulating circadian rhythms. Internal body clock is regulated by the superchiasmatic nucleus (SCN), a small cluster of nerve cells near hypothalamus in the brain. Information of light and darkness from the eyes are conveyed through the Retinohypothalamic tract to the SCN, which synchronizes our body's rhythms to a 24-hour cycle (Moore-Ede 1993).

When homeostatic mechanism and circadian mechanism coincide late at night and in predawn hours, alertness level decreases dangerously. Sleep inertia is the short-term grogginess after awakening.

In addition to these three factors that cause fatigue, there are cumulative fatigue, sub-standard sleep quality and early report time. Cumulative fatigue builds up by sleep debt accumulated in the most recent 24-hour period. Pre-existing sleep debt is carried over from preceding days of inadequate sleep. Adequate sleep requires 7 to 8 consecutive hours of sleep for most people. To recover from the cumulative fatigue, conflicting results of studies show that it takes a several days to three weeks (Nishino 2017). Sub-standard sleep quality is caused by environmental factors, circadian mechanism, sleep disorders and psychological stresses. Sleep onboard is often disturbed by noises, pitching and rolling, a narrow bed, and the coming and going of the crew. Timing of sleep period also affects its quality. Early report time is, for instance, getting out of bed at 03:00 in the morning.

Seafarers work onboard in adverse circumstances with those factors. During night work, the body is forcibly directed to stay awake and to sleep during the day. Studies (Caldwell and Caldwell 2003) show that sleep duration between 11:00 and 19:00 lasts less than five hours. Night work can trigger a bad cycle for cumulative fatigue. After three or four hours of sleep, most night workers awaken, and find it hard to sleep during the day. They engage in normal daytime activities, and report to work at night. There is no room for them to recover from the cumulative fatigue unless taking measures.

Grounding of the *EXXON VALDEZ* was the first case wherein an accident investigation authority determined fatigue to be the probable cause of the marine accident (NTSB 1990). The accident was spotlighted by media on the master's impairment from alcohol, but here the focuses are fatigue and its implications.

Oil tanker *EXXON VALDEZ*

The course of accident and contributing factors

The *EXXON VALDEZ* departed from Trans Alaska Pipeline terminal at 21:12 on 23 March 1989. The pilot took conn until 23:24, then the master took it and decided to depart from the traffic separation scheme in order to avoid glacial ice that had drifted from Columbia bay, and brought the vessel to 180°. After setting the course, the automatic pilot was engaged. The master directed the third mate to return the vessel to the traffic lanes when Busby Is. was abeam, and left the bridge to send messages in his room.

The third mate observed vessel position at 0.9 miles from Busby Is. light, but plotted it 1.1 miles from the light on a chart. A lookout reported to him the Bligh Reef No. 6 light. The third mate directed a helmsman to take starboard 10° after plotting the ship's position but did not confirm it by rudder angle indicator. The third mate called the master to tell that he had started turning the vessel back toward the lanes, and the conversation continued for 1.5 minutes. The lookout reported to him the Bligh Reef No. 6 light again. The third mate checked Reef Is. and Bligh Reef by radar, knew the heading did not change, and ordered starboard 20°. That indicated that the third mate did not notice that his rudder order had not been effective for about 6 minutes. Hastily the third mate ordered hard starboard and notified the master that "we are in serious trouble." At the end of the conversation with the master the third mate felt the vessel grounding at 00:09 on 24 March.

There were deviations from common practices observed in the third mate as follows: (1) not confirming rudder position by rudder angle indicator; (2) not responding to or confirming the reports given by the lookout; (3) communicating with the master for 1.5 minutes while taking conn in the narrow channel; (4) not noticing ineffective rudder order for about 6 minutes.

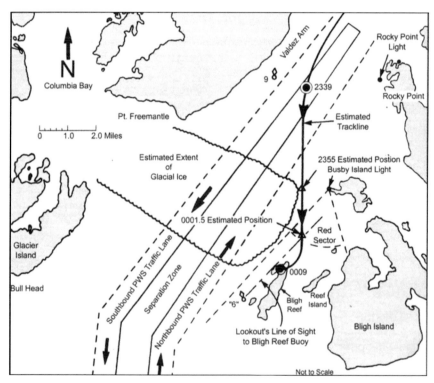

Fig. 4.1: Track of the *EXXON VALDEZ* (NTSB 1990).

The work and rest conditions of the third mate were investigated by NTSB and the following facts were found. After the vessel arrived at berth 5, Alyeska Marine Terminal, at 23:35 on 22 March, the third mate talked with the chief mate about loading crude oil. During loading operation, the third mate and the second mate shared navigation watch duty of the chief mate on 6-hours on, and 6-hours off basis. Transfer of the vessel's ballast water to the terminal started at 00:54 on 23 March. After this, the third mate went to bridge for navigation watch, relieved the chief mate for the meal, assisted him with completing the cargo loading, checked the navigation gear and tested the navigation equipment on the bridge. The pilot and the master got onboard at 20:30. The third mate was relieved by the chief mate and on the stern for standby aft mooring lines at 20:40.

NTSB found that the third mate had only four hours of rest time and one hour of cat nap during the periods from docking the berth to the grounding.

Recommendations concerning fatigue issued by NTSB were as follows.

(1) Eliminate personnel policies that encourage crewmembers to work long hours.

(2) Implement manning policies that prevent long working hours for crewmembers during cargo operations.

(3) Establish a written policy forbidding deck officers to share navigation and cargo watch duties on a 6-hours-on, 6-hours-off basis.

(4) Require two licensed watch officers to be present to take the conn and navigate vessels in Prince William Sound.

Consequences of the fatigue were huge. The *EXXON VALDEZ* spilt an estimated 42 million liters of crude oil, covered more than 26,000 km^2 of water in Prince William Sound and the Gulf of Alaska, and contaminated more than 1,900 km of the coastline. The owner of Exxon paid $1 billion in settlements to the state and federal governments, $300 million in voluntary settlements with private parties, and $507.5 million to more than 32,000 fishermen, native Alaskans and landowners (U.S. Geological Survey 2014).

Sleepiness and accidents

One of symptoms of fatigue is sleepiness. There are many cases of sleepiness related to grounding, collisions and contact.

The *LEVODIJK*, Netherlands flagged container vessel of 9,994 gross tons, proceeding at 17 knots off western Akashi channel, contacted with the seawall of Akashi channel bridge at 04:39 on 19 August 2011. Akashi channel, which leads to port of Kobe, Osaka, is one of the most congested waters in Japan (JTSB 2014a). Navigation watch was done single-handed by the second officer. He had felt sleepiness and then fallen asleep for about 56 minutes until he woke up falling from the chair due to the impact of the collision, and immediately knew that the vessel had made contact with the seawall. In the past he had never slept while on navigation watch and believed he would never fall asleep even if sitting on a chair, and he sat on it.

Adverse effect caused by sleepiness do not often appear in an ocean where vessel traffic is sparse, but are apparent as an accident in the bays and inland seas where vessels, small islands and dangers such as rocks and shallow waters are numerous. Statistics taken by the Marine Accident Inquiry Agency (MAIA 2004) show that most operators believed they would not fall asleep before they sat in a chair and fell asleep. Fifty-nine cases of non-ocean-going vessels involving groundings and contacts caused by sleep during March 2000 and March 2003 that occurred at the Inland waters or Seto-naikai in Japan were analyzed and the result was as follows (Table 4.1).

Concerning the posture at the time of sleep, 90 percent were sitting and 10 percent were standing. All accidents occurred during a single-handed navigation watch, and night watches, 20:24 hours and 00:04 hours, occupy 68 percent of the total number of the accidents.

Statistics on groundings issued by MAIB (MAIB 2004) were similar to the data shown above. Among 66 accidents involving 75 vessels between 1994 and 2003, groundings occupied 31 percent, 23 vessels, and 11 occurred between 00:00 and 06:00. A third of all the groundings of merchant vessels

Table 4.1: Reasons of falling asleep.

Reasons of falling asleep given by operators	%
I never thought I would fall asleep	35
I thought I could stay awake if I had a little patience	21
I thought I would be tolerable because of shift coming soon	17
I thought there was no problem because I took measures	16
I thought there was no problem because I was watchful with meeting vessels	7
I had a cold medicine	2
I did not feel sleepiness	2

over 500 gross tons involved a fatigued single-handed navigation officer at night.

Accidents involving fishing boats caused by operator's sleepiness are abundant. Three hundred forty-seven cases of accidents occurred in Japan between 2008 to 2017 (JTSB 2017a), most of them are grounding, contact with breakwater and collisions. There are special circumstances in which operators and crewmembers tend to overwork by themselves in order to catch more fish on the fishing ground. The situation surrounding fishmen is the same situation not only in Japan but also in other nations.

MAIB issued fishing boat accident flyer in the wake of grounding and sinking of stern trawler *Brothers BF 138* with two crew onboard in 2007 (MAIB 2007). It cautioned that (1) the fishing boat grounded due to one of the crew falling asleep in the wheelhouse, (2) both crew would have been suffering the effect of fatigue brought on by a number of long days at work with only short, broken sleep periods, (3) both crew had also drunk some alcohol before the boat left the harbor, and (4) prawn fishing is seasonal and dependent on good weather. When the conditions for fishing are good, fishermen work very long days to make the most of it. It is not unusual for fishmen to work several 20-hour days.

Effect of fatigue and fatigue countermeasures

Case studies and statistics in previous sections clarify some clues for typical effects of fatigue as follows: (1) Accuracy and timing degrade; (2) Attention decreases; (3) Involuntary lapses into sleep begins to occur; (4) Lower standards of performance unconsciously become acceptable; (5) The ability to integrate information is lost; (6) The ability to maintain a clear picture of the overall situation decreases; (7) Social interaction declines.

Impaired alertness comes from physiological limitations that humans cannot overcome through training or motivational incentives. There are, however, some effective measures for managing alertness or fatigue in night shift workers and sleep restricted workers (Caldwell and Caldwell 2003).

(1) Reduce mental demands on night shifts.

It is apparent that alertness deteriorates between 01:00 and 06:00.

(2) Encourage to double-check everything on night shifts.

It is also part of countermeasures related to item (1).

(3) Make good use of circadian mechanism when taking break.

It is highly recommended that schedule of the break or napping should be selected at low point of alertness of the day: Post-lunch dip, the predawn hours, late night hours.

(4) Take strategic napping against sleep deprivation.

Napping is sleep and most effective when consecutive long sleep is not possible. Even 20 to 30 minute short naps can enhance the productivity of sleep-deprived workers. To be effective, napping should be close to the beginning of long duty work, or be in appropriate time which agrees with the internal body clock between 01:00 and 06:00, and between 14:00 and 16:00.

(5) Try to have a restful night or day of sleep.

When circumstance changes, people often cannot sleep as well as they do at home. To create the same environment onboard as staying at home, for instance, bring own pillow, a family picture or other good reminder of home. Using sleep masks is effective to block out light. Wearing foam earplugs can minimize disturbances and noise.

(6) Consider circadian adaptation.

When shift continues for more than three days, use forward or clockwise rotations. The plan is consistent with physiology that humans have the natural ability to lengthen their internal day easily. If shift work continues only for one to three days, remain on the original daily schedule of eating and sleeping habit at home. In a short period of time, the biological clock cannot adapt to the environmental clock. Remaining on original schedule is better than trying to readjust the internal body clock, which causes fatigue and sleep deprivation.

Fatigue is a real problem in marine sectors, and IMO circulated guidelines for managing fatigue (IMO 2001), which states that good quality of sleep has a positive effect on performance.

4.2.3 Psychological factors

Psychological factors deal with an individual's knowledge and experience with the task and mental capabilities associated with work stations. Mental capabilities include arousal, workload and assumption.

Experience and recency of sailing in the water during night or day are often related to accidents. It includes not only navigational waters, but type of vessel, equipment or instruments and procedures. Arousal is the reaction to stress, biological drive and motivational influences; it ranges from low to high. Arousal level is associated with performance, as shown by the Yerkes-Dodson law. At low level, attention tends to be poorly focused and drifts to task-irrelevant information. At high level, attention tends to be narrowly focused and people neglect peripheral but task-relevant information (Reason and Hobbs 2003). High workload can cause disorganization, fixation, stress and panic, while low workload can cause boredom, inattention and complacency.

In marine accidents, assumption is distinct from other psychological factors. The next section looks into the characteristics of assumption along with its implication to the collisions.

Assumptions and collision

Assumption is an intelligent application of experience based on the belief that things will work the way they are supposed to work, and that people will do what they say they will do. Some seafarers apply it when they meet other vessels. Assumption appeared reasonable to operators at the time of accident for reasons that (1) seafarers will not take needless risks; and (2) they know that things are not always true but are true most of the time (Parrot 2011). This reasoning can be verified by studies on marine accident investigation reports.

A study (Fukuoka 2015) on 28 cases of collisions between 2008 and 2014 that occurred in and around Japanese waters show that among 52 inadequate conditions of operators, phycological conditions accounted for 89 percent in central liveware and most of them were assumptions (Table 4.2). Operators mean officers of watch, masters and pilots who took conns at the time of accidents.

Statistics clarify that operators' assumptions tend to intervene in the process of collision avoidance action and lead to deviation from rules: The Convention on the International Regulations for Preventing Collisions at Sea, 1972 (COLREGs).

Collision between container vessel *KOTA DUTA* and cargo vessel *TANYA KARPINSKAYA* investigated by author clarifies the influence of central liveware in human factors and how the assumption builds up on the accident (JTSB 2014b).

Table 4.2: Pattern of assumptions in collisions.

Pattern of assumptions	Number
Another vessel would take collision avoidance action or alter the course and/or reduce the speed	14
Own vessel would pass starboard to starboard, port to port, etc., to another vessel	7
Another vessel would not approach to own vessel	5
Another vessel would take certain route	4
Own vessel would overtake another vessel safely	3
Approaching vessel would not exist	3
Others	6
Total number of assumptions	42

Collision between the *KOTA DUTA* and the *TANYA KARPINSKAYA*

Summary: The accident occurred at passage crossing at 16:22 on 7 February 2012 when the *KOTA DUTA,* Singapore flag 6,245 gross tons vessel, had left West Wharf No. 3 quay at the Port of Niigata Higashi Ku, Japan, and been sailing toward the Port of Tomakomai, Hokkaido, while the *TANYA KARPINSKAYA*, Russia flagged 2,163 gross tons vessel, was proceeding toward the South Wharf Quay after leaving the Central Wharf quay in the same port. The accident took place just 10 minutes after the *KOTA DUTA* let go of the mooring line (Fig. 4.2).

The course of accident and contributing factors

The master of the *KOTA DUTA* (hereinafter referred to as "master A") sighted the *TANYA KARPINSKAYA* and learned that they would approach at passage crossing. The master A and the master of the *TANYA KARPINSKAYA* (hereinafter referred to as "master B") could not see each other until the distance between the two vessels reached at 890 m because vision was obstructed by buildings on the Steel Yard quay located along the corner of the passage crossing. The master A believed that the *TANYA KARPINSKAYA* was the give-way vessel and requested the ex-master A, who was onboard the *KOTA DUTA* for familiarization, to inquire about the intention of *TANYA KARPINSKAYA* via VHF in order to learn what actions the approaching vessel was going to take.

At first the master A was planning to pass on port-to-port with the *TANYA KARPINSKAYA* at passage crossing. The master B offered to pass

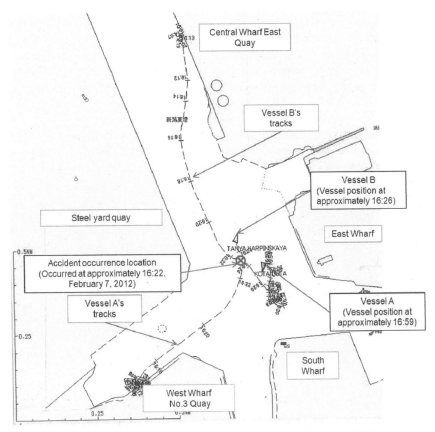

Fig. 4.2: Track of the *KOTA DUTA* and *TANYA KARPINSKAYA* (JTSB 2014b).

on starboard-to-starboard. The master A and the master B agreed to the conduct of vessel to pass on starboard-to-starboard at 16:20 when the distance between the two vessels was approximately 600 m. It took about 20 seconds for both masters to decide while vessels were closing in at 4 to 5 knots.

The master A agreed with offer by the master B because the master B reconfirmed to pass on starboard-to-starboard and ex-master C with a strong tone without a doubt reported starboard-to-starboard. The master A believed that the *KOTA DUTA* could safety pass the *TANYA KARPINSKAYA* if it put the helm hard to port from his maneuvering experience of a sister vessel and that the *TANYA KARPINSKAYA* would alter the course to port to at least 20° to 30°.

The master B offered it because he believed it was easy for the *TANYA KARPINSKAYA* to turn to port due to the large safe area of waters, and safer if two vessel courses had not crossed. He had not enough distance and time to judge the conduct of vessel when he was called via VHF. The master B ordered helmsman to put the helm to port 15°. He thought the hull would be listing and it would be dangerous if putting the helm hard to port since the vessel was fully loaded with iron scraps.

Then the master A ordered hard-a-port to execute the agreed conduct of vessel. Even for a while he was not able to recognize the change of heading of the *TANYA KARPINSKAYA*, whose course changed from 154° to 153°.

The master B also could not see the change of heading of the *KOTA DUTA*, whose heading remained 052° and unchanged after the agreement. He could not understand why the *KOTA DUTA* was not changing the heading to port. He asked the master A through VHF, saying "Have you started turning?" He had not had sea experiences as a master with no more than the size of the *TANYA KARPINSKAYA*: Maneuvering characteristics, especially turning circle, are different depending on a vessel's size.

Both vessels closed in approximately 150 m, both masters felt the danger of collision. At last they took actions to avoid collision, but it was too late to avert it.

VHF communication by VDR readout

Table 4.3 shows VHF communication between the *KOTA DUTA* and the *TANYA KARPINSKAYA* which starts from acknowledging the *TANYA KARPINSKAYA* by Ex-master A onboard the *KOTA DUTA* up to just before the collision. Words written in Italics means the communication was made in Russian language (JTSB 2014b).

Surrounding information

The items to be described next are the surrounding information.

(1) The passage crossing had 560 m width and 14 m depth. It was cloudy with visibility 4 km to 10 km and wind force 2 to 3. There was no strong tidal current.

(2) The master A and the master B had a lot of experience as captain. They did not feel fatigue. The communication via VHF was done in Russian because the master B and ex-master A were Russian. The master A was Polish and learned Russian language which was not his mother tongue. English was used between the master A and ex-master A.

Table 4.3: VHF communication between *Kota Duta* and *Tanya Karpinskaya* (JTSB 2014b).

Time (h:m:s)	Master A	Ex-master A	Master B
16:18:21		A small ship is coming.	
16:18:27	She is moving very slower ...		
16:18:37		...This is Russian vessel.	
16:18:43		Yes, this is Russian vessel.	
16:18:51	Slow ahead.		
16:19:00	Midships.		
16:19:24		Tanya Karpinskaya	
16:19:25	Tanya was ..., why she was coming here.		
16:19:33		Tatiana Karpinskaya, Kota Duta.	
16:19:35			*Kota Duta, Karpinskaya. We'll turn a little to port; let's pass starboard to starboard. We are going to the Lumber wharf (South wharf).*
16:19:44		*Do you want to pass starboard to starboard? Wait!*	
16:19:47			*Yes, we are going to the Lumber wharf. Pass starboard to starboard! Turn now, and we will turn port.*
16:19:48		She wants to starboard to starboard, she wants.	
16:19:51	Starboard to starboard?		
16:19:52		Yes, she makes to port, so you attention.	
16:19:56	OK...		
16:19:58		*OK, you make, turn to port and we will pass starboard to starboard.*	

Table 4.3 contd. ...

...Table 4.3 contd.

Time (h:m:s)	Master A	Ex-master A	Master B
16:19:59	Hard port		
16:20:03			*Yes, we are turning to port.*
16:20:05		*OK, we will turn to port and pass starboard to starboard.*	
16:20:09		She wants to starboard to starboard.	
16:20:18	Dead slow ahead.		
16:20:24		Very fast.	
16:20:26		Cannot, cannot, can damage.	
16:20:28			*Have you started turning?*
16:20:30	Midships.		
16:20:32	You tell him that port to port.		
16:20:35		*Port to port! We can only pass port to port! It's impossible to pass starboard to starboard.*	
16:20:38			*We have already begun turning to port, so we will pull hard to port and turn to port. You turn to port, too. Let's pass starboard to starboard; it's too late (to change).*
16:20:42	Bow, full to port ...		
16:20:44		Yah.	
16:20:45	Stop engine.		
16:20:47		Bow, full to port.	
16:20:48	Bow, full to port, yes.		
16:20:54	Stop engine, already? (Master A received a report "No working, sir." immediately after this from someone other than Ex-master A).		

Table 4.3 contd. ...

...Table 4.3 contd.

Time (h:m:s)	Master A	Ex-master A	Master B
16:20:54			*I am stopping and reversing.*
16:20:59		She is moving to astern.	
16:21:04		Full astern, full astern.	
16:21:07	Full astern.		
16:21:12			*I am stopping and reversing.*
16:21:13		Full astern.	
16:21:15	Astern? Working astern?		
16:21:19		Wow, wow.	
16:21:26	Full astern.		
16:21:31			*Kota Duta! I am reversing; you turn hard to starboard.*
16:21:37		Wow.	

(3) Both vessels were operating the radar but did not properly use it until the master A and the master B saw another vessel with their naked eyes. There were no other vessels that would affect ship maneuvering.

Analysis of information process

Figure 4.3 shows how the master A processed the information about relation with the *TANYA KARPINSKAYA* starting from perception up to taking action to avert collision, as well as how the assumption on his part was building up. The model explains that the master A's situation awareness was perfect, but during the process of decision-making he was influenced by peripheral liveware: Ex-master A and master B. During the process of feedback, he realized the difference between what he expected was going to happen and what was going on in relation to the *TANYA KARPINSKAYA*. He expected that the *TANYA KARPINSKAYA* would steer to port at least 20° to 30°. His expectation was reasonable based on the ordinary practice of seafarer that an operator would take such action in this special circumstance. His expectation had not been challenged to verify by asking the clear messages and commitment of the master B. In

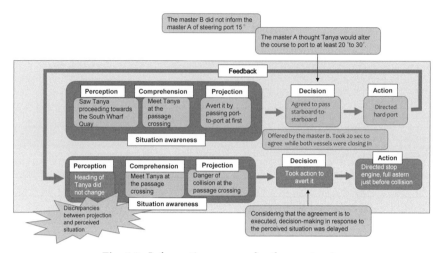

Fig. 4.3: Information process by the master A.

fact, verification is almost impossible in this circumstance when the VHF was used as the communication media among them.

The problems on using the VHF is discussed in central liveware-peripheral liveware interaction: VHF assisted collision.

Woods (1984) studied error detection rate by operators in 99 simulated emergency scenarios tested in a nuclear power plant. Although half of the execution errors were detected by the operators themselves, all mistakes made by operators were not detected by themselves. Those mistakes were detected by a third party. Assumption is categorized as a mistake since there is no discrepancy between action and intention. As a result, assumption is detected or corrected only by fresh eyes.

Parrott (2011) stresses good watchkeeping that requires being alert to discrepancies between reasonable assumption and new information coming in. At the time of accident, five crewmembers—the master A, ex-master A, third mate, cadet and helmsman—were standing on navigation watch of the *KOTA DUTA*; three crewmembers, the master B, chief mate, and helmsman, on the *TANYA KARPINSKAYA*. There were chances for the bridge team to give advice or make an assertion to each master. However, the advice should be appropriate. In this case, it is a common practice taken by the pilots at the port of Niigata Higashi-ku when two vessels are expected to meet at the passage crossing, one of the vessels decreases the speed or stop until another vessel passes through the water.

4.2.4 *Psychosocial factors*

Psychosocial factors include events or conditions in the individual's social environment and are influential when they can manifest physically in the form of stress, anxiety, loss of sleep, fatigue, and so forth.

It is difficult to measure how much psychological factors like interpersonal conflicts, financial problems and lifestyle change can affect individual performance. Holmes and Rahe (1967) provide a measure of the impact of common stressors using life change index scale, based on the observation of important life changes. According to this scale, death of spouse, divorce and marital separation come top three among 43 life changes.

4.3 Liveware-hardware

Liveware-hardware interface is the first of the components which requires matching with the characteristics of the human. It represents the relationship between the human and the machine and includes design and condition of workstations, displays, controls, seats, and all other physical parts of a ship or system.

With advances in technology, the design and material composition of ships have been improving, and machines have been replacing what people had done by hand. Seafarers are expected to monitor more displays than ever before. The human comprehends the information collected by displays from hardware and uses controls to operate the system in accordance with the information. The human sensory capacity is enormous but information transmitting rate during comprehension, projection and decision-making is limited. Hawkins (1987) states that hardware should be designed to fit the human characteristics: Physical, visual, aural, tactical aspects and thought processes.

Grandjean (1969) shows the importance of arrangement in liveware-hardware interface by studying the positional relationship between controls and displays tested on kitchen stoves. In the study, the design of kitchen stoves were divided into four types. The control and the stove burner arrangement were matched only for one type, and three other types were changed in these arrangements. One thousand and two hundred trials were done, no maneuvering error was made in one type that the arrangement matched. Other types lead human errors between the numbers 76 and 129.

IMO (2000a) sets guidelines on bridge layout, and equipment on the bridge, providing ergonomic design of the optimum location for displays, controls and other tools. As for the display and control arrangement, the most important and/or frequently used displays should be located within the central 30° of the viewer's preferred viewing area. The most important and frequently used controls should have the most favorable position with respect to ease of reaching and grasping. The displays and controls should be fitted in a logical arrangement and combined into function groups.

Cranes, piled up containers and other structures on the foredeck can obstruct continuous visual scan by operators and lookouts. IMO clearly regulates that the safe lookout from the navigating and maneuvering workstation should not be influenced by blind sectors. No blind sector caused by cargo, cargo gear or other obstructions outside of the wheelhouse forward of the beam which obstructs the view of the sea surface as seen from the navigating and maneuvering workstation should exceed 10°. The total arc of blind sectors should not exceed 20°. The clear sector between two blind sectors should be at least 5°. Over an arc from right ahead to at least 10° on each side, each individual blind sector should not exceed 5°. Regarding minimum field of vision, the view of the sea surface from the workstation should not be obscured by more than two ship lengths or 500 m forward of the bow to 10° on either side under all conditions of draft, trim and deck cargo.

4.3.1 ECDIS and human factors

In maritime industry, the Electronic Chart Display and Information Systems (ECDIS) became mandatory to almost all large merchant vessels and passenger vessels in July 2018. The ECDIS was introduced in order to enhance safety, reducing in particular the number of accidents involving grounding.

More than 20 types of ECDIS with different displays, controls and functions have been produced. Even ECDIS models from four different manufactures have those critical issues (MAIB 2017a).

(1) The methods of defining the shape and area of a guard zone are varied,

(2) The setting of safety depth and safety contour was inconsistent,

(3) Regarding the guard zone, different labelling, such as look-ahead, safety region, safety zone, safety frame and search light, were used.

Those situations could induce human error, like the aviation field had experienced. In aviation, the cockpit is filled with devices, and the history of improving liveware-hardware interface verifies the decreasing number of human errors in maneuvering those devices.

In addition to liveware-hardware interface, the ECDIS has been triggering liveware-software interface mismatch. This mismatch became apparent by accidents of grounding of vessels equipped with ECDIS. The investigation reports issued by the MAIB (2008a, 2012, 2014a) concluded as follows.

(1) Despite the ECDIS being used as a primary means of navigation, none of the ship's officers had been trained in its use,

(2) the master's and other watchkeepers' knowledge of the ECDIS was insufficient, and

(3) the passage plan was prepared by an inexperienced and unsupervised junior officer and the plan was not checked by the master before departure or by the officer of the watch at the start of his watch.

The ECDIS can trigger another mismatch. Central liveware-peripheral liveware interface mismatch comes from the fact that generic and type-specific ECDIS training is limited to navigation officers. In contrast, education and training on radar have been so widespread through all deck personnel in accordance with STCW Convention that a bridge team can share information and situation awareness.

As the automation of ship maneuvering progresses, there is increasing concern that mismatches of liveware-hardware, liveware-software and central liveware-peripheral liveware interfaces would increase because configuration of a bridge changes to such a situation that is surrounded by navigational and autonomous equipment which is produced by different designs and functions and used by seafarers with varying levels of education and training.

The next accident investigation report (MAIB 2017a) clarifies the implication that ECDIS has on human factors.

Bulk carrier *MUROS*

Summary: The *MUROS*, Spain flagged 2,998 gross tons bulk carrier loaded with fertilizer, was proceeding from Teesport, UK, to Rochefort, France, and ran aground on Haisborough Sand on the east coast of the UK at 02:48 on 3 December 2016. The ship's draft was 6.0 m at forward and 6.2 m at aft. The depth of Haisborough Sand was less than 5 m and the height of tide was 1.2 m.

The course of accident and contributing factors

Before grounding, the second officer (2/O) took over the conn from the master who instructed her to amend the passage plan to route via the Sunk traffic separation scheme (TSS) instead of via the North Hinder Junction. The bridge navigation watch during 00:00 to 04:00 was the 2/O and an able seaman. The master checked and signed the original passage plan prepared by the 2/O before departure from Teesport. The master reviewed and changed it during his navigation watch because he was familiar with the route via the Sunk TSS, which was shorter than the route via the North Hinder.

The 2/O, certified with both generic and type-specific training, was responsible for preparing the passage plans, and got accustomed to using the ECDIS equipped onboard. The 2/O amended the passage plan on the ECDIS during her navigation watch, completed her visual check of the revised route, saved it and printed a copy of the plan, and put it at the rear of the bridge. The ECDIS had the function of "check route" to find hazards in the whole route from berth-to-berth and individual legs of a passage plan. The 2/O knew the function worked when saving the plan but assumed that the hazards along the route were concentrated in the pilotage areas, not confirming them. The revised plan was not checked by the master afterward.

While following the revised track at about 10 knots, at 02:20 the 2/O noticed the ship's speed shown on the ECDIS display had reduced to 9.1 knots and then the speed quickly reduced. The master was called over to the bridge and realized that the vessel was aground when the ECDIS display was zoomed in and the chart view was changed from "standard" to "all", which displayed spot depths, depth contours, deeper than the safety contour, etc.

Contributing factors

Passage planning consists of four stages: Appraisal, planning, execution and monitoring (Swift 2000, IMO 2000b). The contributing factors are analyzed per pertinent stage.

Planning stage

(1) After amending the tack drawn over Haisborough Sand on ECDIS, the 2/O checked it by her visual scan but could not find any danger of grounding. As a result, the 2/O held assumption that track over

Haisborough Sand would be clear by a small safety margin. The 2/O failed to follow the principle of passage planning: The ships always remain in safe water, sufficiently far off from any danger to minimize the possibility of grounding in the event of a machinery breakdown or navigational error. This factor is related to liveware-software interface mismatch.

(2) During post-accident examination on the ECDIS, "check route" page listed over 3,000 warnings from Teesport to Rochefort, including the risk of grounding on Haisborough Sand. The 2/O did not routinely use this function because of irrelevance and triviality of many dangers listed. The number of warnings exceeded the capability of human information process. This factor is related to liveware-hardware interface mismatch.

(3) When using zoom in and "all" mode, spot depth of shallow waters less than one meter on Haisborough Sand appeared on the display, but not when using "standard" mode. This difference came from a technical issue when the Electronic Navigational Chart (ENC) cell was used on this function. The "standard" mode used the smaller scale chart cell, scale 1:700,000, and the "all" mode used the larger scale chart cell, scale 1:90,000, which displayed more accurate and more detailed information. However, the 2/O either did not see the difference or did not recognize the functional implication of the ENC. This event is similar to a mode error. The complex system often employs multi-function controls, and an operator who is unfamiliar with or does not acknowledge the distinction among various control functions can initiate the control incorrectly and lead to an unanticipated system response (Norman 1988). This factor is related to liveware-hardware interface mismatch.

(4) The 2/O amended the passage plan soon after she took over the navigation watch from the master who left the bridge at 00:10. She adjusted the vessel's heading to the revised route at 00:25. She revised the plan by EDCIS in less than 15 minutes. She did not have enough time to confirm that the revised track was clear of danger. This factor is related to central liveware.

(5) The master did not check and approve the revised passage plan made by the 2/O, as required by the SMS manual. The master heavily relied on the 2/O's competence in the use of ECDIS. This factor is related to liveware-software interface mismatch.

Monitoring stage

(1) The ECDIS could generate alarms related to navigation safety, such as shallow waters, but the audible alarm had been disabled. The ECDIS was a type of model wherein the guard zone had to be active for the alarm to be triggered. In this case, the guard zone was not ticked, resulting in it being inactive. These settings were protected by passwords, and crew considered these inactive setting to be beneficial. The accident report did not clarify the crew's decision, but it is considered that they inactivated the alarm because frequent sounds and alarms, which are not necessarily issuing critical information, could distract their attention and impact on their performance. This factor is related to liveware-hardware interface mismatch.

(2) The 2/O did not recognize shallow waters, less than 5 m in depth, which were marked by a buoy on Haisborough Sand. The intended track apparently crossed the water between the one marked by the buoy and Haisborough Sand. The bridge watch during 00:00 to 04:00 coincides with the lowest alertness level in the circadian rhythm. The 2/O was sitting in a chair for two uneventful hours on the navigation watch. The accident investigation report concludes that the 2/O's performance was possibly affected both by circadian mechanism during the time of day and by low level of arousal caused by sitting in a comfortable chair. These factors are related to central livewire.

(3) The bridge team consisted of the 2/O and lookout at the time of grounding. There was no report on the buoy at Haisborough Sand or the fact that the vessel was heading to shallow waters by the lookout. The lookout was not required by the STCW Convention to be certified with either generic or type-specific training on the ECDIS. Sharing mental model among the bridge team cannot be expected as long as the lookout's unfamiliarity with the ECDIS exists. This factor is related to central liveware-peripheral liveware interface mismatch.

4.4 Liveware-software

Liveware-software is the second interface with which human factors are concerned. It reflects the relationship between the individual and supporting systems, or non-physical part of the system found in the workplace. It includes regulations, manuals, checklist layout, charts, maps, publications, standard operating procedures, computer software design and the safety management system.

Ready availability, usage and contents of manuals, checklists, and other documentations are related. Documentations include essential qualification and certification for the task. Those documentations need to incorporate mandatory rules and regulations, applicable codes, guidelines and standards recommended by flag states, classification societies and maritime industry organizations.

Reason (2008) found that in nuclear industry about 70 percentage of all human performance problems were related to procedures. These procedures include wrong information, inappropriate and unworkable situation, out-of-date information and absence of related task, and these procedures were not found in workplace nor by workers.

Reason's findings are useful for considering procedures which have flaws or deficiencies as contributing factors in human error. When all people across the world use agreed procedures, it is important to think about which stages of the procedures are related to the accidents and incidents. When operators observe the movement of another approaching vessel at sea, they take collision avoidance actions in accordance with COLREGs. The next study clarifies which stage of collision avoidance process stipulated in the COLREGs is related to the accident.

4.4.1 Collisions and COLREGs

Fukuoka (2015) studied 28 collision cases of serious marine accidents investigated between 2008 and 2014 by JTSB, in which each stage of the procedures was looked into in order to find factors contributing to the collision. The study included 54 vessels, excluding fishing boats and pleasure boats. Collision avoidance procedures are stipulated in COLREGs and compared to risk management process (ISO 2009, IEC 2009).

Rule 5 covers risk identification; Rules 7(a), (b), and (c) cover risk analysis; Rule 7(d) covers risk evaluation; Rules 8(a) and (b) cover risk treatment; Rule 8(d) covers monitoring and review. In the case of restricted visibility, the first paragraph of Rule 19(d) covers risk analysis and Rules 19(d)(i) and (ii) and 19(e) cover risk treatment.

The pertinent sections of COLREGs are as follows

(1) Rule 5 states that every vessel must always maintain a proper lookout to enable a full appraisal of any collision risk.

(2) Rules 7(a), (b), and (c) state that every vessel must use all available means to determine whether any collision risk exists. Radar equipment, including long-range scanning, must be used properly to obtain early

warnings of any collision risk and to undertake systematic observation of detected objects. Assumptions must not be made on the basis of insufficient information.

(3) Rule 7(d) states that a collision risk exists if the compass bearing of an approaching vessel does not change appreciably.

(4) Rules 8(a) and (b) state that any collision avoidance action must be taken in accordance with this rule and in ample time with good seamanship. Any alteration of course and/or speed to avoid collision must be large enough to be apparent to the other vessel.

(5) Rule 8(d) states that the collision avoidance action must result in passing at a safe distance. The effectiveness of the action must be carefully monitored until the other vessel is past and clear.

(6) The first paragraph of Rule 19(d) states that in the case of restricted visibility, a vessel that detects the presence of another vessel by radar alone must determine whether a close-quarters situation is developing, and/or any collision risk exists.

(7) Rules 19(d)(i) and (ii) state that when a vessel detects the presence of another vessel by radar alone, it must not alter its course to port if the other vessel is forward of the beam except to overtake the vessel; moreover, it must not alter its course toward a vessel abeam or abaft.

(8) Rule 19(e) also states that every vessel must reduce its speed to the minimum at which it can maintain its course when it receives the fog signal of another vessel forward of its beam or cannot avoid a close-quarters situation with another vessel forward of its beam.

Table 4.4 shows that unsatisfied risk analysis by an operator, i.e., OOW, master or pilot, at the time of an accident contributes to most collisions, regardless of daytime and nighttime. The findings indicate that most operators did not use all available means to determine if risk of collision existed, and they made assumptions on the basis of insufficient information.

Risk identification was the second leading unsatisfied risk management process, which meant operators did not always maintain a proper lookout to make a full appraisal of the risk of collision, and they did not notice other approaching vessels until accidents occurred. However, there was no big difference in percentage between daytime and nighttime. From a different viewpoint, operators can use radar by day and night to detect approaching vessels with the same accuracy.

Table 4.4: Unsatisfied risk management process observed at 54 vessels in collisions.

Locations	Daytime		Nighttime		All day	
	Number	%	Number	%	Number	%
Risk identification	4	18	6	19	10	19
Risk analysis	16	72	22	69	38	70
Risk evaluation	1	5	0	0	1	2
Risk treatment	1	5	4	12	5	9
Monitoring and review	0	0	0	0	0	0
Total number of vessels	22	100	32	100	54	100

A big difference appears in risk treatment process, which means that collision avoidance action is more difficult at night than in the daytime. It is considered that an operator's situation awareness is related to the fact that most of critical information about a target vessel is obtained by radar at night. Relationship between radar image and collision avoidance action is discussed in section "Restricted visibility and collision avoidance action" in Liveware-environment interaction. Furthermore, avoidance action by an operator in close quarter situations during the night is not the same as in the daytime. It is common practice that operators do not use radar in close quarter situations due to limits derived from radar performance.

4.4.2 Collision and passage planning

The procedure of passage planning stipulated in IMO guidelines is utilized among seafarers as well as COLREGs. Analysis of the grounding of the *MUROS* indicated that the intended track on the ECDIS could lead to grounding. There are cases wherein an intended track itself holds the risk leading to the collision.

Collision between YONG SHENG VII and HOKUEI No. 18

Summary: The *YONG SHENG VII,* Panama flagged 2,982 gross tons cargo vessel, and the *HOKUEI No. 18,* Japan flagged 960 gross tons dredger carrier, collided in the fairway of Port of Kin-nakagusuku, Okinawa in Japan at 1919 on 15 November 2014. There were no casualties, but *HOKUEI No. 18* flooded after the collision and rolled-over to the starboard side, grounding to the seabed (JTSB 2016a).

The course of accident and contributing factors

Figure 4.4 illustrates intended tracks and AIS tracks of both vessels. The master of the *YONG SHENG VII* (hereinafter referred to "master A") planned to proceed in the middle of a 13 m dredged fairway if there was no meeting vessel, while the master of the *HOKUEI No. 18* (hereinafter referred to "master B") planned to navigate in the middle of lines of buoyed waterways. The master A saw an approaching inbound vessel the *HOKUEI No. 18* with navigation lights just after leaving the wharf, thought that the *HOKUEI No. 18* would navigate in the middle of dredged fairway, and decided to keep their ship near to the outer limit of the lines of buoyed waterways.

The master B, single-handed at the bridge, could not notice the approaching the *YONG SHENG VII* until the last minute. The master observed radar before passing near the breakwater but could not identify any outbound vessel. On this side, there were communication issues between the master B and chief officer stationed at the bow.

Both masters had different recognition for which waters, i.e., lines of buoyed waterways or dredged fairway, they should navigate. There had been no clear message about it from the port authority. Passage plans involved serious risks of collision at the onset.

Rule 9 in COLREGs states that a vessel must keep as near to the outer limit of the channel or fairway which lies on her starboard side as is safe and practicable. A fairway is a navigable waterway for a large vessel, usually marked by a pecked line on a chart, while a narrow channel is marked by lines of buoys. It is not appropriate for a vessel to navigate it by the right side only after meeting a vessel proceeding in the opposite direction. A vessel is expected to proceed to keep near to the outer limit of the right side (Cockcroft and Lameijer 1982).

4.4.3 Casualties during mooring operations

Even if the mooring system design is not changed, a tool box meeting and risk assessment should be encouraged for crewmembers in order to identify any hazard prior to and during the mooring operation, which includes assessing the ship's movement followed by adjusting a mooring line, observing other crewmembers standing within a safer area, looking at abrasion damage to mooring lines, improper spooling of wire lines, both mooring wire and ropes going through the same Panama lead, etc.

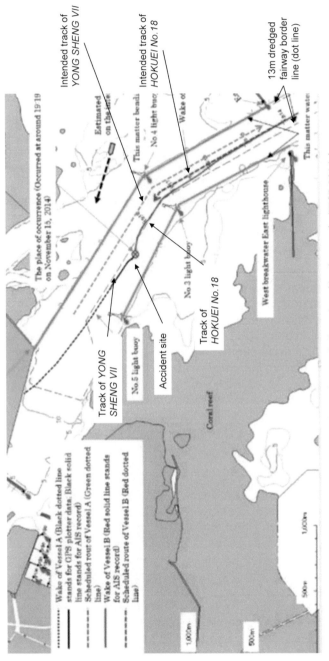

Fig. 4.4: AIS tracks and intended tracks of both vessels (JTSB 2016a).

Color version at the end of the book

The first case study raises issues already acknowledged by IMO circulated guidance, shown above on snap-back zone, and mooring arrangement or mooring system design itself. The second case illustrates the importance of risk assessment prior to adjusting a mooring line.

LNG carrier *ZARGA*

Summary: While the *ZARGA*, Marshall Island flagged 163,922 gross tons liquefied natural gas (LNG) carrier fully loaded with LNG, was under a berthing operation at the South Hook LNG terminal, Milford Haven, UK, the high modulus polyethylene (HMPE) mooring rope that was used as a fore spring line parted and struck the officer-in-charge (OiC) of the forward mooring party, resulting in serious injury at 19:08 on 2 March 2015 (MAIB 2017b).

The course of accident and contributing factors

On 28 February, the master held a pre-arrival planning meeting on the mooring plan, attended by the C/O, OiC of the forward and aft mooring parties and the bosun. They agreed to use 10 mooring lines attached fore and aft, respectively, in accordance with the standard procedure for South Hook. On 1 March the master, senior officers and bosun discussed a detailed work plan and made instructions which direct nine mooring lines in a 3-4-2 configuration fore and aft. The instructions were posted on the crew noticeboard and required the bosun to circulate it in a toolbox talk. OiCs did not attend the meeting, therefore, when the *ZARGA* arrived at the entrance of the port at 15:00, the OiC of the aft mooring party noticed that the laid-out mooring lines did not agree with the pre-arrival planning meeting, he directed the team to add a tenth mooring line and notified the bridge to ready for arrival.

At 15:10, two pilots boarded the *ZARGA* and instructed 10 mooring lines fore and aft, considering the weather. Nine mooring lines at bow station were laid out and were not altered. The *ZARGA* was maneuvered port side alongside the South Hook jetty, assisted by tugs. At 16:25 the first fore spring line was connected, and at 17:23 three breast lines fore and aft were connected, respectively.

At 17:53 the OiC of the forward mooring party informed the master that two forward head lines had been made fast. The master replied that there should be three head lines. The OiC reported the master that one of the head lines had been used as a breast line and all the breast lines needed to be repositioned. The C/O was sent forward to solve the problem. The master, the C/O, the pilots and the shore mooring team discussed the most appropriate and quickest ways of changing the forward mooring arrangement for 15 minutes.

Up until then, lack of communication or break down of communication loop between the master and the OiC had contributed to the mooring lines problem even though a toolbox talk and briefing prior to the mooring operation were conducted.

The OiC had been rearranging the mooring lines, the C/O returned to the Cargo Control Room (CCR), and the OiC of the aft mooring party started to refit rope chafing guards by slackening off the aft spring lines. At 18:28 the OiC of the forward mooring party reported that the forward lines were in position. The pilot directed the tugs to stop pushing the *ZARGA* onto the berth and let go. At the forward mooring station, the mooring party began to refit chafing guards to the ropes.

At 18:54 the C/O was informed by the terminal staff that the *ZARGA* had moved forward and was out of its designated position by 1.5 m, beyond the normal tolerance. The master decided to move the vessel aft by using the forward spring lines. As the forward mooring party was already released, the OiC directed the bosun to take control of the mooring winch. The C/O monitored the ship's position at CCR.

The OiC did not call the re-muster of the forward mooring party, as a result he got involved in part of the mooring work that other crew could do instead of supervising the whole forward mooring operation. The situation increased the likelihood of accidents during mooring operations.

Under the direction of the master, the aft spring lines were slacked, then fore spring lines were heaved up. The bosun heaved up the fore spring lines until the winch stalled under the load, released the control lever, and operated it again. By 19:05 misalignment of the vessel was corrected to 75 cm. At 19:08 the 3/O saw the forward inboard spring line parted and then noticed the OiC lying on the deck, near the spring line roller fairleads.

Snap-back zone

The OiC of the forward mooring party had been standing in the area aft of the spring line ship's side roller fairleads, which was designated as a safe zone in accordance with the ship's guideline provided by the ship management company. The guideline stated that crew are not in danger of snap back if they are not standing between the mooring winch and the fairlead at the ship side. The guideline was reflected on the *ZARGA*'s snap-back plans and as a result giving a wrong perception among crew that standing outside the painted snap-back zone would be safe.

Code of Safe Working Practices for Merchant Seamen (COSWP) states the limitation of traditional snap-back assessment methods and design complexity of mooring decks on modern ships, and recommend discarding previous snap-back zone conception and taking an entire mooring deck to be regarded as being dangerous.

MAIB report states, concerning the snap-back zone in this case, that an accurate snap-back zone could not have been predicted without practical test-based data and computer modeling software; however, more thorough snap-back assessments would have indicated that the area where the OiC was standing was a potential snap-back zone. Basically, a parted rope recoils along its connected line, for instance from one fairlead to another fairlead. If the broken end strikes an object during recoiling it can be changed to another direction. Consequently, the recoil trajectory creates a new danger zone and the potential snap-back zone becomes more complex.

Other contributing factors

(1) HMPE ropes fitted on board the *ZARGA* in replacement of steel wire ropes were originally used in offshore sectors after the sector's extensive testing of Working Load Limits, Minimum Breaking Strength, etc., through engineering perspective. When using HMPE ropes as steel wire ropes at ship-shore mooring operations, verification in engineering perspective is necessary.

(2) Environmental conditions like strong westerly winds and tidal range should have been considered and the master should have called the once released tugs to help secure vessel position because more than half of the mooring line failures occurred between 2010 and 2015 at South Hook and Isle of Grain LNG terminals. The master, with the consultation of the pilot, decided to adjust the exceeding 1.5 m forward ship's position by manipulating the spring lines. The situation lead crew on the mooring station to be put in a more hazardous situation.

Cargo vessel *ONOE*

While the *ONOE*, Japan flagged 87,404 gross tons bulk carrier, was loading sea salt alongside the starboard side of the pier of the Morro Redondo Port in the Cedros Island, Mexico, a wire rope which supporting the shore gangway parted and the 2/O standing on the gangway was thrown off, resulting in death between 03:45 and 03:55 on 17 December 2013 (JTSB 2015a).

The course of accident and contributing factors

After the arrival at the Morro Redondo Port on 14 December 2013, the first officer (1/O) had a meeting with port manager on the loading cargo works where they agreed that the shore gangway operation was to be conducted by crewmembers of the *ONOE*, and that gangway was to remain at a higher position than the ship's upper deck handrail and separated from the vessel at a safe enough distance when it was not used. The 1/O notified all crewmembers of the agreement, as well as through "Night Order Book".

At 03:45 on 17 December, the 2/O, who was working in the cargo control room as 00–04 watch with an able seaman and ordinary seaman (OS), was called by one of two stevedores to accompany them and measure the draft of the vessel from the work barge waiting alongside the pier. The 2/O walked toward the pier through the shore gangway, and then the 2/O noticed aft mooring lines were slacked. He instructed the able seaman to wake up the bosun and another crewmember for the next watch duty. He directed the OS to heave up slacked aft mooring lines with the mooring winch while standing on the shore gangway which had been set at the same level of the handrail.

After having asked the 2/O's permission for the mooring operation, the OS started to heave up the mooring lines. While heaving up the mooring lines, the gangway contacted with the handrail, then moved toward the shore direction. There was nobody left around the gangway and the height of the gangway could not be adjusted. The wire rope which supported the gangway gave way and the 2/O was thrown from the gangway to the pier, about 5 to 6 m height (Fig. 4.5).

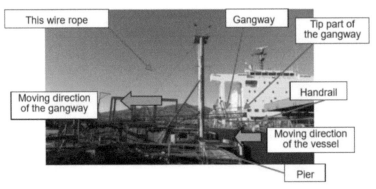

Fig. 4.5: Ship movement and the shore gangway (JTSB 2015a).

The ship management company had established procedures which prescribed that the mooring winch operation was recommended to be conducted in accordance with this manual. The manual that involved safe mooring practices and risk management prescribed the following: (1) Crew adjusting the mooring lines need to obtain permission from the Officer of the Watch (OOW). The OOW need to carefully assess how the position of the vessel would be changed by the adjustment of the mooring rope prior to giving permission; (2) Crew engaging in the mooring operation are required to be trained so that hazard will be identified, and risk assessment will be implemented prior to the mooring operation.

Ship management company did not provide education and training about the risk assessment for all the crewmembers of the *ONOE*. The 2/O did not conduct the risk assessment prior to giving the OS permission to use the winch.

4.5 Liveware-environment

The Liveware-environment interface represents the relationship between the individual and the internal and external environments. Internal environment includes noise, temperature, light, vibration, motion, air quality, and the atmosphere that people work in. External environment includes weather, visibility, sea conditions, traffic density, waypoint, bottle neck, ice condition, regulatory climate and the broad political and economic constraints under which the marine system operates.

Parrott (2011) states that transition is a period in which a current condition changes from one state to another in environment and ship's operation. Transition can lower the level of the operator's situation awareness. Transition occurs in such cases as: (1) changes of watch, (2) change of crew and leadership, (3) a pilot onboard, (4) change of ship management company and other organization which affects ship's operations, (5) changes of passage plan, (6) chart changes, (7) changes in settings, e.g., VHF radio channel, (8) changes of equipment, e.g., change from paper charts to ECDIS, or change of a different type of ECDIS itself, (9) twilight, (10) visibility, (11) bottlenecks, and (12) waypoint and waterways.

From the viewpoint of liveware-environment interface and implication that has an impact on shipping industry, twilight, visibility, bottlenecks, waypoints, a narrow channel, sea and weather conditions, regulatory climate, and economic constraints are discussed.

4.5.1 Twilight

Here, more attention is paid to dawn than dusk. During twilight, in which depth and color perception become weak, operators experience difficulty in distinguishing objects and lights. In addition, dawn coincides with circadian trough, the human's lowest level of alertness. As a result, physiological factors of liveware and adverse environment for operators can work against their performance.

4.5.2 Visibility

When it changes from good visibility to restricted visibility, depth and color perception become as weak as during twilight. Operators have to rely on radar images and fog signals sounded by vessels in the vicinity of their own vessel. Not all vessels are detected by radar; small fishing boats and leisure boats, including yachts made with fiber reinforced plastics (FRP) or woods, are often not identified if radar reflector or AIS are not equipped. Sea conditions with high waves, heavy rain and snow can become a factor of losing situation awareness among operators who do not know how to adjust radar with varying control.

Collision between NIKKEI TIGER and HORIEI-MARU

Collision between the *NIKKEI TIGER*, Panama flagged 25,074 gross tons bulk carrier, and the *HORIEI-MARU*, Japan flagged 119 gross tons fishing vessel, is a case wherein the 2/O could not detect the *HORIEI-MARU* by radar during adverse weather conditions. It occurred at 01:56 on 24 September 2014 off the coast of northern Japan in the Pacific Ocean. A helmsman of the *NIKKEI TIGER* reported to the 2/O a masthead light of a fishing vessel on the port bow more than 30 minutes prior to the collision. With about 3 m wave height, wind force 7 and rain, the 2/O changed the radar range to try to detect the fishing vessel but could not. Relying on the navigation light displayed by the fishing vessel, the 2/O took collision avoidance actions and ordered the helmsman to take hard port. After collision, the *HORIEI-MARU* sank, and nine crewmembers were rescued but 13 were missing (JTSB 2014c).

4.5.3 Restricted visibility and collision

Relationship between radar image and collision avoidance action

MAIA studied collision avoidance actions in 91 cases of collision in restricted visibility or fog that involved 157 vessels, including cargo vessels, tankers, and passenger vessels, that occurred between 2001 and 2005 (MAIA 2007). In restricted visibility conditions, situation awareness mostly relies upon the information coming from the radar. This study explains how the operators respond to the information from the radar in order to avoid collision.

There were 74 vessels that did not reduce the speed to the minimum or take the way off while acknowledging the development of a close-quarter situation after the initial recognition of an approaching vessel. The avoidance actions taken by these vessels were as follows: (1) 17 vessels altered the course to port (23 percent); (2) 27 vessels proceeded straight ahead (36 percent); and (3) 30 vessels altered the course to starboard (41 percent).

Among the 17 vessels which altered the course to port, 12 vessels detected the image of the other vessel on the right side of the ship's heading marker (SHM) (71 percent). There were two vessels that detected the image of the other vessel on the left side of the SHM (12 percent), and whose reasons were that there were other vessels nearby on the starboard side proceeding in the same direction or the opposite direction. The other three vessels detected the image of the other vessel on the SHM, and their reasons were that their own vessels were approaching the shoreline on the starboard side (17 percent).

Among the same 17 vessels shown above, 16 of the opponent vessels that were involved in collisions altered the course to starboard, and only one vessel proceeded straight ahead. Considering the conduct of vessels in restricted visibility prescribed in COLREGs, the finding makes it clear how dangerous it is for the vessel to alter the course to port in order to take a collision avoidance action in restricted visibility, resulting in a collision. Table 4.5 shows the reasons of altering the course to port in 12 vessels which detected the other vessel's image on the right side of the SHM.

A contributing factor of altering the course to port on observing the opponent vessel's image on the right side of the SHM can be explained in Fig. 4.6. It illustrates the actual course and relative course of the opponent vessel on the radar display. When observing the image in relative mode, it appears that the heading of the opponent vessel has a small angle with

Table 4.5: Reasons for altering the course to port in restricted visibility among 12 vessels.

Reasons for altering the course to port	Number of vessels
An operator intended to increase the passing distance between the opponent vessel in ample time	5
There were other vessels on starboard side nearby proceeding the same direction or the opposite direction	4
Just the location of opponent vessel's radar image was at the right side of SHM	2
An operator assumed the opponent vessel would alter the course to port	1

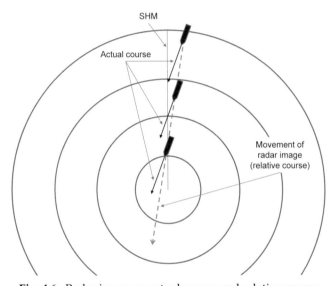

Fig. 4.6: Radar image on actual course and relative course.

own SHM, as a result, it appears that enough room would be given if own vessel altered the course to port at a small rudder angle. When using relative mode, actual course of the opponent vessel is further towards the portside of own vessel. Furthermore, small course alteration is not recognized as any course change by the other vessel on radar display.

Therefore, it is encouraged to use "Trial Maneuver" and closely monitor Closest Point of Approach (CPA) and Time of CPA (TCPA) functions if fitted in radar.

4.5.4 Bottlenecks and waypoints

Bottlenecks here mean geographical and bathymetric change of safe navigation waters, such as going from open water to harbors, bays, sounds and vice versa. In these waters, vessel traffic converges, currents and tides affect vessels' speed and maneuvering. Turning too late or early can have serious consequences and overtaking and meeting situation can increase the risk of collision in these waters. More often do they become good fishing grounds where there are a number of various types of fishing boats with fishing gears drawn from the stern or side that could influence the intended course of the merchant vessels. In addition, during night, various kinds of colors and rhythms emitted by lights in accordance with the Maritime Buoyage System keep navigators busy sorting out the critical information about safe navigation during the passage. When visibility is restricted by fog, passage of these waters might be suspended by local regulations. This measure is considered reasonable on view point of central liveware itself and liveware-environment interface in order to prevent accidents.

Figures 4.7 and 4.8 show accidents data investigated and issued by JTSB between 2008 and 2017 on Japan Marine Accident Risk and Safety Information System.

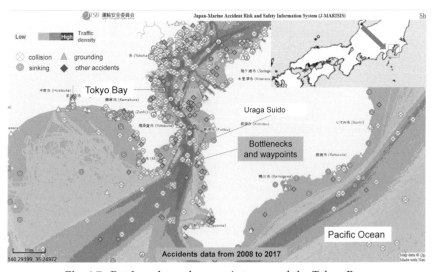

Fig. 4.7: Bottlenecks and waypoints around the Tokyo Bay.

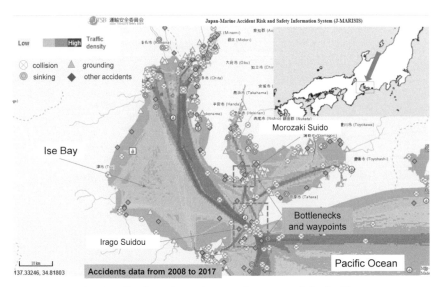

Fig. 4.8: Bottlenecks and waypoints around the Ise Bay.

The Tokyo Bay and the Ise Bay

The Tokyo Bay and the Ise Bay are the most congested waters of marine transportation in the world. The data in Figs. 4.7 and 4.8 add other findings to common knowledge that heavy traffic density is associated with an increased number of marine accidents, such as collisions. This is clear when looking at the data of accident location and heavy traffic density on the route between the Tokyo Bay and the Pacific Ocean as well as between the Ise Bay and the Pacific Ocean. More accidents, except in and near ports and harbors, occur in the vicinity of waypoints at bottlenecks. The next two collision cases show how accidents occurred in these waters.

Collision between *BEAGLE III* and *PEGASUS PRIME*

Summary: A collision between the *BEAGLE III*, Panama flagged 12,630 gross tons cargo vessel, and the *PEGASUS PRIME*, Korea flagged 7,406 gross tons container vessel, occurred at Uraga Suido, Japan at 03:10 on 18 March 2014. Seven crews of the *BEAGLE III* died and two are missing among the 14 crewmembers onboard, and the ship sank after collision. One crew of the *PEGASUS PRIME* was injured, and the ship's bow buckled (JTSB 2016b).

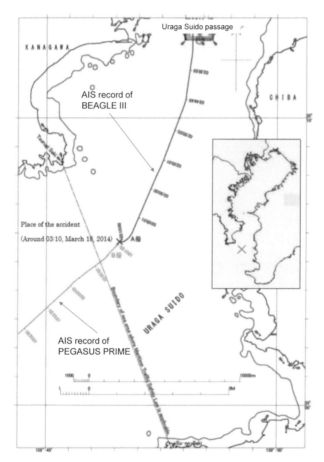

Fig. 4.9: AIS tracks of BEAGLE III and PEGASUS PRIME (JTSB 2016b).

These are courses of maneuvering taken by both officers on the watch (OOW) at the time of accidents. The OOW of the *BEAGLE III* died, so the evidences were drawn from AIS records and the witness from a lookout (Fig. 4.9).

The course of accident and contributing factors

Maneuvering of *BEAGLE III*

After the *BEAGLE III* navigated out of the south exit of the Uraga Suido passage, the master left the bridge at 02:45, and the 2/O and an able seaman took charge of the navigation watch. The 2/O ordered starboard 10° and then noticed the starboard sidelight of the *PEGASUS PRIME* ahead at 03:07. He could not have noticed the *PEGASUS PRIME* on the starboard bow before the course alteration because the *BEAGLE III* turned to starboard about one minute after the *BEAGLE III* turned to port. The 2/O took collision avoidance action by taking helm hard to starboard at 03:09, which was estimated by the fact of acceleration of turning angular velocity to the starboard side.

Maneuvering of *PEGASUS PRIME*

The 2/O and an able seaman took charge of the navigation watch, the *PEGASUS PRIME* proceeding toward a way point off Turugizaki. The 2/O recognized the *BEAGLE III* at four miles port ahead by radar and a starboard sidelight by binoculars and thought that the *BEAGLE III* would alter the course to avoid own vessel since both vessels were in crossing situation. When the *BEAGLE III* approaching two miles port ahead and did not change the course to take collision avoidance action, the 2/O ordered own vessel to turn to port. In the distance of 0.5 miles starboard ahead, the 2/O noticed the *BEAGLE II* was turning to starboard, thought it was impossible to avoid a collision, and directed the able seaman to make port 20° in order to mitigate the damage. The 2/O did not have a way to know which direction or port the *BEAGLE III* planned to navigate towards after leaving Uraga Suido passage.

Four stages in a collision situation

Although Rule 17 of COLREGs does not mention the distance between the two vessels approaching in a crossing situation in sight of each other and the corresponding actions taken by them, there may be four stages at open waters (Cockroft and Lameijer 1982).

(1) More than five to eight miles away, both vessels can take any action.

(2) Rule 17(a)(i) starts to apply to vessels of which the distance at which they are approaching each other is between five to eight miles. A give-way vessel must keep out of the way of the stand-on vessel. Rule 15 makes it clear that the give-way vessel must avoid crossing ahead of the stand-on vessel, in other words, it must alter course to port. The stand-on vessel must keep her course and speed.

(3) Rule 17(a)(ii) applies when the distance is between two to three miles. The stand-on vessel can take action to avoid collision by maneuvering

alone when it becomes apparent that the give-way vessel is not taking appropriate action by CORLEGs. The stand-on vessel is required to give whistle signals of at least short five blasts, prescribed in Rule 34(d), and must not alter course to port, considering the give-way vessel may take action to alter simultaneously or subsequently.

(4) When the give-way vessel is so close that collision cannot be avoided by their action, the stand-on vessel must take action as will best aid to avoid collision.

When the *BEAGLE III* was approaching two miles port ahead, the 2/O of the *PEGASUS PRIME* ordered to alter course to port, and then found the *BEAGLE III* was turning to starboard. The main reason of the collision becomes clear comparing OOW's maneuvering with the four stages at open water shown above.

Environment on accident site

According to the sailing direction issued by Japan Coast Guard in 2014, average passage number of vessels a day at Uraga Suido was about 400 vessels in 2011. In Tokyo bay, traffic is quite heavy with all types of vessels, and, in addition, a large density of fishing boats may be encountered in the vicinity of passage.

Figure 4.10 shows the volume of marine traffic per hour for 24 hours prior to the collision time of this accident at Uraga Suido. The number of the vessels reflects the vessels equipped with AIS; therefore, other vessels that were not required to equip, such as small coastal vessels and fishing boats, are not included in the number. The total number of vessels was 302; there were 150 north-bound vessels heading toward Uraga Suido passage and 157 south-bound vessels leaving Uraga Suido passage.

Figure 4.11 illustrates AIS tracks of the vessels corresponding to Fig. 4.10, and locations of similar accidents. There were four cases of collision at Uraga Suido as follows: 04:42 on 2 September 1997; 18:30 on 7 October 1998; 18:10 on 25 January 2000; and 05:18 on 13 April 2006. Those accidents occurred either during the hours when the number of north-bound vessels was the greatest or south-bound vessels was the greatest during the time of the day.

Uraga Suido becomes waters of waypoints for vessels to proceed to and from three different directions: Vessels proceeding southwest after leaving Uraga Suido passage; vessels coming from western waters to enter Uraga Suido passage; vessels coming from eastern waters to enter Uraga Suido passage. In other words, convergence and divergence of marine traffic

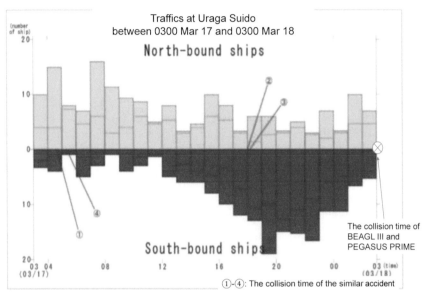

Fig. 4.10: Marine traffic around the accident site (JTSB 2016b).

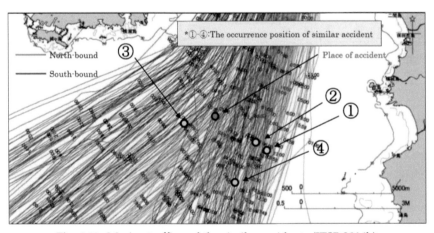

Fig. 4.11: Marine traffic and the similar accidents (JTSB 2016b).

Color version at the end of the book

with numerous vessels occurs at Uraga Suido at the same time on a daily basis. In this environment, operators are put under a difficult situation in which they are forced to address uncertain issues of target vessel's intention in a limited time span while considering meeting situations with other vessels in the vicinity at the same time and in restricted sea-room.

Although the accident investigation report does not mention the possibility of being affected by other vessels which were navigating near both the *BEAGLE III* and the *PEGASUS PRIME* on their maneuvering in the crossing situation, uncertainty of other vessels might have influenced the course of action taken by both OOWs based on the presence of heavy marine traffic, shown in Fig. 4.10, and marine traffic around the time of this accident and similar accidents shown in Fig. 4.11.

The next accident case, occurring in Ise Bay, makes clear the reason of and the course of uncertainty.

Collision between *DAIO DISCOVERY* and *AURORA SAPPHIRE*

Summary: A collision between the *DAIO DISCOVERY*, Panama flagged 40,245 gross tons cargo vessel, and the *AURORA SAPPHIRE*, Philippine flagged 28,074 gross tons cargo vessel, occurred near Irago Suidou in Ise Bay, Japan at 19:22 on 13 March 2011. There were no casualties, but both vessels suffered minor damages on their port side (JTSB 2011a).

Both vessels were conned by pilots; The *DAIO DISCOVERY* passed the Irago Suido passage and was heading north to the Port of Kinuura, while the *AURORA SAPPHIRE* departed from the Port of Kinuura and was heading South towards the Irago Suidou passage via Morozaki Suidou (Fig. 4.12).

The course of accident and contributing factors

Maneuvering of *DAIO DISCOVERY*

The pilot of the *DAIO DISCOVERY* (hereinafter referred to "pilot A") saw navigation lights of the *AURORA SAPPHIRE* while turning No. 3 light buoy after leaving the Irago Suidou passage and thought the *AURORA SAPPHIRE* would proceed westward to circumnavigate the Okinose light buoy, indicating shallow waters, not proceed South directly to the Irago Suido passage. The pilot A believed that the circumnavigation route was common knowledge among pilots in stormy weather, more than 4 m wave height was observed at the entrance of the Irago Suido passage, although the circumnavigation route was not clearly defined by procedures of the Pilot Association.

Assuming that the *AURORA SAPPHIRE* would take the circumnavigation route, much attention of the pilot A was paid to the

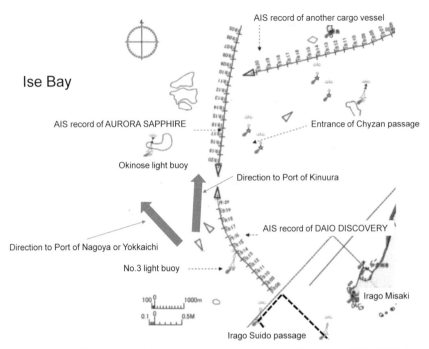

Fig. 4.12: AIS tracks and situation surrounding the accident site (JTSB 2011a).

two other approaching vessels. One was a cargo vessel meeting with near intended course off the Chyuzan passage. He made contact with a pilot onboard another cargo vessel by VHF and confirmed to avoid a crossing situation. Another one was a fishing boat with a white light lit proceeding port bow in an overtaking situation. The pilot A ordered to turn the No. 3 light buoy by 10° rudder angle, considering safety margin with the fishing boat. The master of the *DAIO DISCOVERY* (hereinafter referred to "master A") noticed the approaching vessel, the *AURORA SAPPHIRE*, and reported it to the pilot A. The maser A received some reply from the pilot A and thought the pilot A had already made contact with the pilot of the *AURORA SAPPHIRE* (hereinafter referred to "pilot B") but did not know the content of communication made between the pilots since it was not made in English language. The pilot A noticed the *AURORA SAPPHIRE* approaching 0.7 to 0.8 miles on the bow when the master A reported that the sidelight of the *AURORA SAPPHIRE* had changed from red to green.

81

Maneuvering of the *AURORA SAPPHIRE*

The pilot B had never experienced such adverse weather conditions wherein the wave height exceeded 4 m and pilot station was changed to the waters near No. 3 light buoy. There were no clear messages from the Pilot Association they belonged to about the navigation route to the No. 3 light buoy. The pilot B did not know the common practice to circumnavigate Okinose; he directed the vessel to proceed south to the pilot station without doubt.

While proceeding south, the pilot B noticed three approaching vessels including the *DAIO DISCOVERY* in the Irago Suido passage. In his experience, after leaving the termination of Irago Suidou passage, 80 percent of the vessels were heading northwest toward the Port of Nagoya or Yokkaichi. He assumed that the *DAIO DISCOVERY* might go northwest, not going north which would develop a meeting situation with his own vessel. He could not see the navigation lights of the *DAIO DISCOVERY* by naked eyes through the windows not equipped with a clear view screen because of a relative wind speed of 20 m/s and the heavy rain. He did not make use of AIS information about the approaching vessels.

Approaching to the pilot station, the pilot B had been talking with the master of a pilot boat accompanied about time of disembarkation from the *AURORA SAPPHIRE*. The pilot B could not notice navigation lights of the *DAIO DISCOVERY* until the master of the *AURORA SAPPHIRE* reported to him that the green sidelight had changed to red in about 500 m on the bow.

Both pilots took collision avoidance and damage mitigation actions.

Contributing factors

Contributing factors of the accident are as follows: (1) as organizational factors, the Pilot Association did not clarify the passage route to the pilot station in case of adverse weather conditions, (2) there was a failure of pilot and bridge team integration, not sharing critical information about approaching vessels, (3) both vessels encountered adverse weather conditions with strong wind and high waves, (4) both pilots held assumptions on an another vessel's intended route, (5) uncertainty for intended routes of vessels, including the *DAIO DISCOVERY* and the *AURORA SAPPHIRE*, was caused both by convergence and divergence of heavy vessel traffic in narrow waters and restricted sea-room forced by heavy traffic, and (6) both pilots had an overload situation, in that they were forced to address more than two meeting situations with other vessels at the same time.

4.5.5 *Narrow channel*

Figure 4.13 illustrates the Kanmon Kaikyo where open waters and restricted inland waters meet, it covers about 15 miles and is known as the difficult place of the voyage. The Kanmon Kaikyo holds seven designated passage and port sections by Japanese laws and consists of important waterways where about 510 vessels passed on a daily basis in 2015.

The data makes it clear that accidents recur in long narrow channels where waypoints coincide. Among them, there were seven collisions which occurred between 2008 and 2017 at the narrowest waters in the channel off Mojizaki, which measures about 650 m in width with maximum 10 knots current. All seven collisions were related to an unexpected course deviation caused by the influence of strong currents while adjusting the helm in order to prevent a close quarter situation with other vessels in an overtaking or head-on situation. In the channel, large angle veering is not recommended.

Accidents also occurred at the intersection of two passages where vessels alter their headings to a new route. All these accidents were related to assumptions made by an operator when analyzing the course that an encountering vessel might take.

The Malacca Strait, where 4,825 very large crude oil carriers (VCCC) transited in 2013, the English Channel, the Dover Strait, where over 400 vessels pass through daily, etc., have some similar geographical and bathymetric characteristics as Tokyo Bay, Ise Bay and Kanmon Kaikyo and feature shallow waters scattered in the channel and strait. When

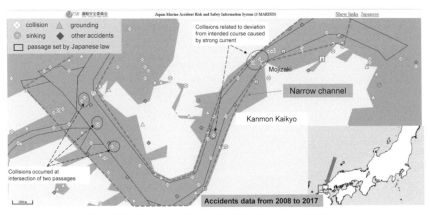

Fig. 4.13: Narrow channel of the Kanmon Kaikyo.

transiting those waters, mariners are required to handle both vessel traffic and navigation at the same time.

4.5.6 Atmosphere where people are working in

During loading and unloading, there have been lots of occupational casualties. According to the statistics, with 9,291 cases of total number of accidents from 2008 to 2017 issued by JTSB, 15 percent of accidents were associated with occupational casualties, such as casualties occurring during loading and mooring operations (Fig. 4.14).

Accidents involving enclosed space entry, where mariners or workers work in different atmospheres, occurred every year. There were three major contributing factors. One was that they did not measure the atmosphere in an enclosed space prior to the entry. Another was that they did measure the atmosphere but did not do it at different levels in spaces like a cargo hold or cargo tank. The last one was that standard procedures on enclosed space entry were not established, and education and training to crewmembers on enclose space entry, including procedures, measurement of atmospheres,

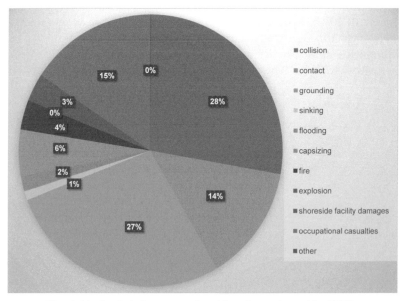

Fig. 4.14: Statistic by types of accident investigated by JTSB.

donning of PPE and preparedness to deal with emergency situations, were not provided by the organizations. Under these circumstances, the worst case can develop into the secondary and tertiary accident, when people without measuring the atmosphere or donning proper PPEs attempt to rescue the victims who became unconscious on the inner bottom plating of cargo holds. Those three factors are entwined with each other.

The next two cases of occupational casualties illustrate the course of accident and contributing factors of why and how people risk their own lives at an enclosed space.

Chemical tanker *KYOKUHOU-MARU No. 2*

Summary: The casualty occurred onboard the *KYOKUHOU-MARU No. 2*, Japan flagged 388 gross tons chemical tanker, during an unloading operation at Kawasaki-ku, Port of Keihin, Japan, at 13:55 on 10 March 2010 (JTSB 2011b). The chief officer (C/O) who entered a cargo tank died of suffocation due to oxygen deficiency.

The course of accident and contributing factors

When loading the tertiary butyl alcohol (TBA) at a previous berth, the C/O forgot to put a drain plug on the cargo piping in the starboard No. 2 cargo tank, and ordered it to be loaded without putting it after it had been removed for an inspection by a surveyor prior to the cargo tank survey. The work of fixing the drain plugs had usually been done by the C/O alone. The *KYOKUHO-MARU No. 2* finished loading and proceeded to the berth for unloading, where the accident occurred.

When arriving at the berth in the Port of Keihin at 11:45 on 10 March, the C/O and a worker confirmed the items written in the cargo handling checklist. They checked that nitrogen gas was planned to inject into all cargo tanks while unloading. Then, unloading commenced and nitrogen gas was injected into the cargo tanks as a return gas during unloading operation in order to prevent explosion and negative pressure in the cargo tanks. Except for the C/O and the chief engineer (C/E), the rest of the crew, including the ship manager, were not made aware of the nitrogen injection.

Almost at the end of unloading, the cargo pump sounded like it was inhaling air, and the TBA was not unloaded. The C/E entered the cargo pump room to investigate any leak from the pump to find why TBA was not transferred from the cargo tank to a shore tank. And then the C/E went to the fore deck, opened the tank hatch of starboard No. 2 cargo tank, and looked inside the cargo tank through the access hatch. He noticed that a drain plug was placed on the stage set in the cargo tank and informed the C/O to that effect.

The C/O looked inside the starboard No. 2 cargo tank through the access hatch and noticed that the drain plug was not fixed on the cargo piping. He entered the cargo tank wearing a gas mask without measuring the atmosphere. When handling TBA, wearing PPE such as a gas mask is required. About 10 seconds after the C/O's attempt to fix the drain plug, the C/E noticed that the body of the C/O was lying face down and stopped moving.

The C/O suffocated due to a lack of oxygen in the cargo tank. The oxygen level in the starboard No. 2 cargo tank measured after the accident, at about 14:23, was 16 percent, the acceptable oxygen level being 21 percent.

Another worker working near the shore tank heard the occurrence of an accident onboard the *KYOKUHOU-MARU No. 2* and went to the pier where he saw three crewmembers wrapping a rope around their waists and preparing to enter the cargo tank. The worker was worried about a possibility of a secondary disaster because he had learned about similar cases during education and training sessions conducted by his company. To prevent the crew from entering the cargo tank, he intervened and recommended that they stay there until an ambulance arrived. The secondary disaster was prevented. At 14:44 the C/O was rescued by a fire department rescue team and transported to the hospital, but his death was confirmed.

There were no procedures regarding enclosed space entry established onboard the *KYOKUHOU-MARU No. 2*. Neither were all crewmembers educated and trained on the danger of enclosed space entry. The ship owner and general safety manager did not prescribe in a manual that access hatches must not be opened during a cargo operation because they thought such information was common knowledge among crewmembers onboard chemical tankers. They could not foresee that a crewmember would enter the cargo tank during the loading and without measuring the atmosphere.

Cargo vessel *SINGAPORE GRACE*

Summary: While the *SINGAPORE GRACE*, Hong Kong flagged 15,071 gross tons bulk/lumber carrier, was berthed at the wharf of Port of Saganoseki for discharging cargo work of copper sulfide concentrate at 08:30 on 13 June 2009, one of the workers (Driver B) fell while descending a ladder for cargo work at No. 3 cargo hold. Two of the three other workers (the foreman and Operator C) who went to rescue him also collapsed in the cargo hold. All three workers were rescued from No. 3 cargo hold, but later they were confirmed dead. At 08:50, at about 10 cm above the cargo surface, where the foreman, Driver B and Operator C had collapsed, the O_2 concentration was 1.5–2.0 percent (JTSB 2012b).

The course of accident and contributing factors

The *SINGAPORE GRACE*, after 12 days of voyage from Port Moresby Harbor in Australia, moored at the private berth in Port of Saganoseki, Japan, at 07:48 on 13 June and opened the hatch cover of cargo hold No. 3 to discharge the copper sulfide concentrate. The foreman measured the O_2 concentrations in cargo hold No. 3 by himself from 07:50 to 08:05.

The foreman instructed the chief officer (C/O) to open the hatch cover and then he posted the access permit notice board at the entrance hatch of cargo hold No. 3 based on a regulation which was created after the fatal accident occurred four years ago.

The foreman did not measure the atmosphere in accordance with the standard procedures set by the stevedoring company he belonged to. According to the form of the record book produced by the stevedoring company, the record had 12 blank spaces to record the O_2 concentration per hold, i.e., six blanks for lower layer measuring points under fore, center and aft hatch coamings of both sides and six blanks for upper, middle and lower measuring points under fore and aft entrance hatches.

Supervisors of cargo operation, including the foreman of the stevedoring company, were accustomed to measuring O_2 concentrations without using the determined methods. The smelter company and the stevedoring company were not aware of these habitual practices used in the measurements of O_2 concentrations by supervisors of cargo operation, including the foreman.

Driver B (heavy vehicle driver) entered cargo hold No. 3 at 08:30 since the access permit notice board was posted at the entrance hatch of the hold and another operator had already started driving a heavy vehicle in cargo hold No. 1. Driver B inhaled oxygen-deficient air and developed anoxia.

The reason for the oxygen-deficient air was as follows. The copper concentrate had oxidized during transportation from Port Moresby Harbor to port of Saganoseki, and the oxygen in air tight cargo hold No. 3 had been consumed, creating an oxygen-deficient environment. Copper sulfide concentrates were beneficiated from copper ore by floatation method using reagents. Depending upon the properties of the floatation reagent adhered to copper sulfide concentrate, it may generate toxic gas. The toxic gas is heavier than air; therefore, it stagnates in cargo holds. As a result, the danger of the toxic gas was not being replaced by air and the atmosphere of the cargo holds became oxygen deficient.

The secondary accident

The foreman, who was informed that Driver B had collapsed, entered the cargo hold No. 3 to rescue Driver B together with Operator C and Operator F (crane operator) without measuring the atmosphere. As a result, the foreman inhaled oxygen-deficient air and developed anoxia, resulting in the secondary accident. The foreman was not aware of oxygen-deficient atmosphere in cargo hold No. 3 for the following reasons.

(1) The foreman felt impatient and responsible to rescue Driver B, and he lost his sense of composure,

(2) There were workers who had a misunderstanding that oxygen-deficient conditions in cargo holds were removed by natural ventilation as time passed after opening the hatch covers,

(3) Measurements for detecting an oxygen-deficient atmosphere had not been done by the time this accident occurred, and

(4) Four years ago, the smelter company and the stevedoring company experienced a fatal accident onboard another vessel during loading operation due to anoxia in a hold, since then, there had been no accidents.

The tertiary accident

Operator C together with Operator F entered cargo hold No. 3 wearing gas masks to rescue the foreman and Driver B, inhaled oxygen-deficient air and developed anoxia.

Operator C entered cargo hold No. 3 once again for the following reasons.

(1) Operator C thought that he could enter an oxygen-deficient atmosphere by equipping with gas masks,

(2) Since Operator C had already developed anoxia when he had gone to rescue Driver B at the time of the primary accident, he could not make appropriate decisions when he entered cargo hold No. 3 once again.

The secondary and the tertiary accident occurred between 08:30 to 08:40. Appropriate education and training on actions to be taken in the case of fatal accidents had not been provided to personnel by the stevedoring company, which contributed to the secondary and the tertiary accident.

Those facts found by the accident investigation demonstrate the importance of measuring the atmosphere in different levels and education and training on how to safely enter the enclosed space.

Oxygen-deficient atmosphere at different layers

O_2 concentration was varied at each layer of the cargo hold No. 3 as follows. (1) O_2 concentration around the upper deck was approximately 19 percent at around 09:07. (2) O_2 concentration at middle layers of the hold from around 08:30 to 08:40 was approximately 12–16 percent. (3) O_2 concentration at lower layers of the hold was approximately 1.5–2.0 percent at about 08:50.

Contributing factors of continuous oxygen-deficient environment

The oxygen-deficient condition lasted for the following factors: (1) Odorous hazardous gases were generated by the floatation reagents and were heavier than air. Therefore, they were not replaced by fresh air and subsequently accumulated in the lower layer of the cargo hold. (2) The air in cargo hold No. 3 was not replaced by outside air through natural ventilation with the wind velocity under 0–1.4 m/s. (3) The smelter company and the stevedoring company did not consider the necessity of forced ventilation because they had experienced that O_2 concentration increased to 20.9 percent by natural ventilation as time passed.

4.5.7 *Sea and weather condition*

Sea and weather conditions affect all types of accidents. They can contribute to listing, capsizing and sinking. They can also lead to groundings, where in some cases vessels were anchoring in and off a harbor during stormy weather, and their anchors were dragged. Other groundings were cause by losing an opportunity for the master to shift the vessel to a safe sheltered anchorage. Dragging the anchor itself is an incident, but grounding followed by dragging anchor is an accident. What are the different factors working in these situations?

First, look at the regulatory requirement of anchor equipment in relation with environmental conditions, and then look into the contributing factors in the case of grounding and capsizing.

Regulatory requirement of anchor equipment

According to the International Association of Classification Societies (IACS) unified rules for the design of anchoring equipment (IACS 2016), the maximum environmental loads are as follows:

(1) Maximum current speed is 2.5 m/s,

(2) maximum wind speed is 25 m/s, and

(3) the anchoring equipment required is intended to be used in a harbor or sheltered area, where there are no waves, not to be used off exposed coasts in stormy weather.

The rules describe the equivalent conditions shown above:

(1) Maximum current speed is 1.5 m/s,

(2) maximum wind speed is 11 m/s, and

(3) maximum wave height is 2 m.

Concerning anchor winch motor performance in the windlass design load, minimum lifting capacity is three lengths of chain, that is 82.5 m and the anchor.

In order to achieve the necessary anchor holding power, the class guidelines recommend that the length of chain deployed is 6 to 10 times the water depth.

Group of shipping industry consisting of Det Norske Veritas-Germanisher Lloyd (DNV-GL), Gard and the Swedish Club (2016) found in the study on anchor losses that the majority of the losses were caused by exceeding the maximum environmental loads under which anchor equipment was used. Locations of losses were outside a harbor or sheltered area.

Limitation of anchoring in significant environment

Mariners are often forced to confront with environmental conditions that exceed those maximum environmental loads. Significant environmental conditions are accompanied by strong low pressure, typhoons, cyclones, hurricanes, etc.

Marine Accident Inquiry Agency (MAIA: present organization name is Japan Marine Accident Tribunal) surveyed and analyzed 825 vessels of more than 100 gross tons on associations between the fact of anchor dragging and different weather conditions that were caused by some of ten typhoons that hit Japan in 2004 (MAIA 2006). The types and number of surveyed vessels were (1) 296 passenger vessels and ferries; (2) 228 oil tankers; (3) 148 chemical tankers; (4) 78 general cargo vessels; (5) 39 Liquid Petroleum Gas vessels; (6) 20 Roll/on roll/off vessels; and (7) 16 car carriers. MAIA verified a formula that is empirically used as a guide for vessels lying at single anchor and found the limitation of anchor equipment under these weather conditions. The guide of the official formula shown below describes the relationship between the water depth and the length of anchor chain when lying on single anchor prescribed in the marine maneuvering course used by the former Japanese navy. In the study, vessels were divided into two categories, taking wind pressure receiving area into consideration: One is passenger vessels and ferries; another is cargo vessels which include from (2) to (7) shown above.

The guide of the official formula

When a vessel receives wind speed 20 m/s blown from bow 30° on each side at a sheltered area, the length of anchor chain which would avoid dragging anchor is:

$3D + 90$ (m)

D is the depth at high water.

When a vessel receives wind speed 30 m/s in stormy weather, blown from bow 30° on each side at a sheltered area, the length of anchor chain which would avoid dragging anchor is:

$4D + 145$ (m)

The guide adds that the most dangerous situation in the stormy weather is anchoring in waters exposed to ocean or waters with strong tides and current, the former causing severe pitching and the latter causing a large angle swing due to inconsistency between the wind direction and tidal current direction, resulting in dragging anchor.

MAIA survey findings

The MAIA study found the following:

(1) $3D + 90$ m was effective for smaller passenger vessels and ferries of less than 1,000 gross tons until wind velocity of 20 m/s and wave height of 1 m. For cargo vessels, the same formula was effective until wind velocity of 20 m/s and wave height of 1.5 m.

(2) $4D + 145$ m was effective for passenger vessels and ferries until wind velocity of 25 m/s and wave height of 2.5 m. For cargo vessels the same formula was effective until wind velocity of 30 m/s and wave height of 2 m.

(3) With these surveys and additional simulation experiments, the study concludes that when lying at single anchor at good holding ground, 3D + 90 m is effective with wind velocity up to 20 m/s, 4D + 145 m is effective with wind velocity up to 30 m/s.

Difference between an incident and an accident in stormy weather

When mariners do not follow the maximum environmental loads prescribed in IACS or the guide of the official formula lying at single anchor, they might face an incident.

In addition to exceeding the maximum environmental loads, when these following situations coincide, grounding is likely to occur.

(1) When selecting an anchorage, not considering that the vessel at the anchorage is exposed to dominant wind and swell and then vulnerable to approach the shallow waters or shoreline.

(2) Not gathering updated weather forecast and losing an opportunity to shift a vessel to safer waters.

(3) Not considering the limit of windlass performance to heaving up anchor in the stormy weather.

(4) Not considering the limit of ship's maneuverability at given drafts to proceed intended course or route up to a shelter in the stormy weather after heaving up anchor.

The next two cases of groundings provide details about the different factors contributing to an accident. In the first case, a vessel was involved in a failure of heaving anchor followed by dragging anchor due to the stormy weather, in the second case, a vessel succeeded heaving anchor but failed to proceed with the intended course to a safe shelter.

Cargo vessel *COOP VENTURE*

Summary: The *COOP VENTURE*, Panama flagged 36,080 gross tons cargo vessel, was riding a single-anchor paid out six shackles at 25 m depth of sand bottom in Shibushi Bay, southern Japan, when a typhoon was approaching. The ship's agent recommended the master to shift to Kagoshima Bay where was a shelter area for large vessels. The master decided not to shift and stay there. When the typhoon approached, the vessel started dragging anchor by wind velocity of 25 m/s and wave height of 5 m, and then run aground on the shallow water at 21:15 on 25 July 2002 (Fig. 4.15). During abandon ship activities, four crew lost their lives, and the vessel became a total loss (MAIA 2003).

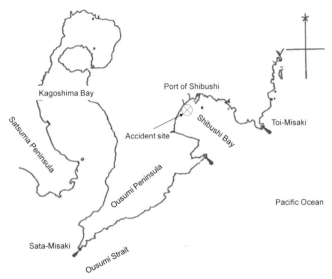

Fig. 4.15: Accident site of grounding (MAIA 2003).

The course of accident and contributing factors

While unloading grain, the agent recommended the master to shelter in Kagoshima Bay because a typhoon was approaching. The *COOP VENTURE,* while loading 40,280 tons of grain with fore draft of 8.0 m and aft draft of 11.6 m, headed outside the Port of Shibushi at 09:10 on 21 July.

There was enough time before the typhoon reached the bay. At this point the master had three choices: Sheltering in Kagoshima Bay, sheltering in open ocean far away from the storm area of the right side, called dangerous semicircle, and anchoring in Shibushi Bay. The master selected the last one as the typhoon was approaching 600 miles east southeast away from the port proceeding west northwest at a speed of 15 knots.

Shibushi Bay is exposed to ocean and subject to high waves from east to south. Sailing direction cautions that anchoring in Shibushi Bay should be avoided when wind is blowing from east to south, sending ocean waves into the bay. According to the weather forecast issued by Japan Meteorological Agency, Shibushi Bay was expected to enter the dangerous semicircle at 19:00 or 20:00 on 25 July.

At 09:00 on 25 July, the master knew that the typhoon was approaching 191 miles east southeast from anchoring position at the speed of 17 knots, leaving the possibility of Shibushi Bay entering the dangerous semicircle.

The master assumed that the force of the typhoon would gradually weaken because the wind speed had not reached 10 m/s, wave height had not been so high as about 2 m, and the class of typhoon had decreased by one level. In addition, it took about 11 hours for the *COOP VENTURE* to arrive in Kagoshima Bay and the route included proceeding south, which meant heading toward the center of typhoon. He considered sheltering in Kagoshima Bay as dangerous and decided to stay here, lying at single anchor with the main engine adjusting for lessening tension on the anchor chain. He had many experiences encountering typhoon, hurricanes and cyclones, and knew the consequences brought by these meteorological phenomena.

The wind suddenly became stronger and it began to exceed 15 m/s, and the wave height of the swell which entered from the bay entrance gradually became higher. At 16:24 the master put a helmsman and used the main engine at dead slow ahead, slow ahead, and stop while observing the direction of the swell. At 19:30, wind blowing from the bay mouth was reaching 17 m/s, the maximum instantaneous wind speed was 28 m/s, and the wave height of the swell reached about 5 m.

At 20:30, the *COOP VENTURE* started dragging anchor when the northeast easterly wind reached 25 m/s, with a maximum instantaneous wind speed of 41 m/s, and the maximum wave height of the swell reached 8 m. The master used the main engine at full ahead, half ahead and so forth to prevent the vessel from drifting to shore. He found it ineffective and decided to heave anchor. When two anchor chains were on deck, windlass could not heave up the remaining four anchor chains. The vessel could not stop drifting and ran aground in water depth of 10 m at 21:15. The hull broke up from amidships, leaking 228 tons of fuel oil which polluted the beach.

MAIA's simulation results

Anchor holding power and tensions

Figures 4.16 and 4.17 show the relationship between holding power and tensions put on the anchor chain by wind and swell. In the simulation conducted by MAIA, the conditions were set up with the same environments that the *COOP VENTURE* encountered: A starboard AC type anchor weighing 6,975 kg, weight of anchor chains at 146 kg per 1 m, water depth of 25 m, wave height of 5 m, and sand bottom.

Figure 4.16 illustrates that, when receiving wind only from the bow, the tension of the anchor chain due to the wind pressure was about 8 tons

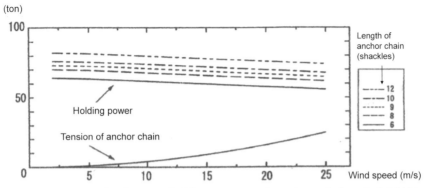

Fig. 4.16: Relationship between holding power and tension only by wind (MAIA 2003).

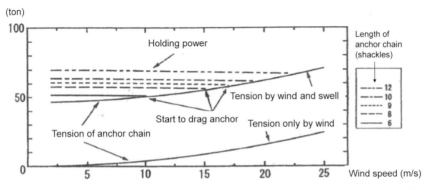

Fig. 4.17: Relationship between holding power and tension by wind and swell (MAIA 2003).

at a wind speed of 15 m/s, and 25 tons at 25 m/s. As the wind pressure increased, the bottom landing portion of the anchor chain decreased, and the six anchor chains had a holding power of about 60 tons at a wind speed of 15 m/s, and 55 tons at 25 m/s. However, the holding power exceeded the wind pressure and could afford sufficient reserve capacity.

Figure 4.17 shows that, when receiving both the wind and wave from the bow, the tension of the anchor chain, due to the wind pressure and the wave drifting force, reached about 50 tons at the wind speed of 10 m/s, and exceeded the anchor holding power of six shackles. With the wind velocity of 25 m/s, they exceeded the anchor holding power of 12 shackles.

These results indicated the swell greatly affected the dragging anchor.

Length of anchor chain and the survey findings

The amount of anchor chain that the *COOP VENTURE* paid out at the water depth of 25 m at the time of accident was six shackles or 165.0 m. When applying this situation to the MAIA survey findings, anchor chain elongation was equivalent to 3D + 90 m, which is effective only for wind velocity up to 20 m/s and wave height of 1.5 m.

If the *COOP VENTURE* had paid out nine shackles or 247.5 m, the elongation would have been equivalent to 4D + 145 m, which is used as a measure in the stormy weather of wind velocity of 30 m/s and wave height of 2.5 m.

As shown in Fig. 4.17, even if the amount of anchor chain is paid out to 12 shackles, the result of dragging anchor will be the same.

Feasibility of using main engine

To prevent dragging anchor, the simulation results showed the necessity of continuous propeller thrust of 8 or 9 knots using the main engine. On a practical viewpoint, MAIA concludes that, in this case, it is difficult to maintain the normal anchoring state for a long time by using the main engine. One reason is that it is necessary to adjust propeller thrust by using various engine motions while observing swing motion of the hull and the state of anchor chain. Another is that it is not easy for large vessels to fine-adjust the thrust force, especially at nighttime when it is hard to confirm the state of ship motions.

Cargo vessel *CITY*

Summary: The *CITY*, Panama flagged 4,359 gross tons cargo vessel, was lying at a single-anchor paid out 6.5 shackles near quarantine anchorage of the Port of Sakata, northern Japan, under ocean wind warnings issued. When the wind velocity became stronger, the master noticed the danger of dragging anchor, and directed to heave up anchor, attempting to shift the anchoring position. After heaving anchor, the vessel was driven by a strong wind pressure and stranded on tetrapod located at the breakwater of the Port at 05:09 on 10 January 2016. There was no fatality, but the vessel became a total loss (JTSB 2017b).

The course of accident and contributing factors

Prior to anchoring, the master had obtained the Asia Pacific ground analysis diagram and the coastal wave analysis figure by weather faxes, he thought that there were no signs of weather deterioration in these analysis charts. When deciding the length of anchor chains, the master considered ability of the vessel to quickly change an anchoring position in case of weather deterioration and 6.5 shackles were paid out at 01:13 on 10 January. The vessel had 10 shackles on each side. The anchorage was 35 m deep and the bottom material was mud mixed with sand.

At 03:00, the vessel began to swing centering around the anchoring position as a result of wind direction changing from south to northwest as the wind blew harder. The master received information about ocean wind warning with 15 m/s maximum wind speed from the Japan Coast Guard, noticed the danger of dragging anchor, and decided to heave up anchor.

Japan Coast Guard, Japan Meteorological Agency and NAVTEX had already issued an ocean wind warning the day before the accident. In addition, sailing directions issued by the Hydrographic Department of the Russian Federation state that 50 to 60 days of storm are observed at the Port of Sakata in a year and 80 percent of them are concentrated between November and March. Vessels entering the port can preferably take refuge in Tobishima, northwest of the Port of Sakata, when the north to northwest wind is strong.

On its aweigh anchor, the master ordered full ahead, trying to sail for offshore at 04:04, but the vessel failed to obtain a speed necessary for keeping the course and ran onto the breakwater under northwest wind velocity of 14 to 17 m/s and wave height of about 2 m.

The vessel was under ballast voyage with all ballast tanks except the forepeak tank filled with water in fore draft of 2.25 m and aft draft of 3.75 m, which meant the vessel was exposed to much more wind pressure. The output power of the vessel's main engine was 2,400 kW which was smaller than the average output power, about 3,226 kW, of most cargo ships of the same level of gross tonnage.

The safety management manual of the vessel's management company did not mention the seaworthiness, such as a limit anchor holding power and limit wind speed in a ballast condition and a limit of ship maneuvering for keeping course considering wind pressure and output power of the main engine.

The master did not pay much attention to information about the predictable sea and weather conditions and appropriate anchorage, thereby missing the opportunity to shift anchorage when the vessel could have still been controlled.

4.5.8 *Regulatory climate and economic constraints*

Rasmussen (1997) pointed out conflicts among organizations after analyzing marine accidents as follows: (1) the strategies for legislation are inadequate against advanced technology; (2) legislators are influenced by the shipping industry; (3) ship owners influence classification societies; (4) design based on established practice becomes obsolete in the fast-pace of technological change; (5) lack of communication between designers, manufacturers and ship operators.

The next accident investigated by MAIB clarifies the external environment implication to the capsize of the cement carrier *CEMFJORD*, all of whose crew were missing (MAIB 2016). This section focuses on external environment; the course of accident and other contributing factors are discussed in the case study in Chapter 10: Convergence of accident models.

Cement carrier *CEMFJORD*

Summary: The *CEMFJORD*, Cyprus flagged 1,850 gross tons cement carrier, fully laden with 2,084 tons of cement, departed a cement loading terminal at Rordal in Denmark on a heading to Runcorn in U.K. on 2 January 2015 with significant safety deficiencies on rescue boat launching arrangements and cargo hold bilge pumping system. The *CEMFJORD* capsized in extraordinarily adverse sea and weather conditions at 13:16 on 2 January while transiting the Pentland Firth, Scotland.

Regulatory climate: Frequent safety exemptions

Ship management company, Brise Bereederungs, planned to replace the *CEMFJORD's* two lifeboats with a rescue boat because of the frequent malfunction of two old lifeboats. The plan was rejected by the flag state. According to the SOLAS regulations, the *CEMFJORD* had to carry a rescue boat or a lifeboat on one side of the ship so as to accommodate all persons on board. The rescue boat it planned to carry had a capacity of six persons, however, the whole crew was eight persons. Brise Bereederungs pressed ahead with the plan and vessel's recognized organization (RO), DNV-GL, approved it applying the guidance by the flag state; however, at this point, the flag state had never approved it.

By 28 November 2014, the *CEMFJORD* was in dock in Poland and had lifeboats replaced with a rescue boat. The DNV-LG surveyor rejected the new rescue boat arrangements due to boat fall failure. Brise Bereederungs asked the flag state consul in Hamburg to update the SOLAS exemption by an email saying that it had decided to install the "new boat" on the starboard side and the port side lifeboat was in good working condition. And then the exemption was issued immediately. It referred to the lifeboats, not the rescue boat arrangements. There was different understanding on the content of exemption among ship management company, RO, and flag state.

(1) Brise Bereederungs understood that the replacement was approved by DNV-GL and flag state consul approved the vessel's operation without rectifying rescue boat launching arrangements.

(2) DNV-GL approved the modification from a technical view point, applying flag state guidance on SOLAS equivalence arrangement.

(3) Flag state consul in Hamburg did not understand the content of exemption when it had been informed by the Brise Bereederungs of modification.

(4) Flag state head office in Limassol did not know of the work undertaken in dock and believed that the *CEMFJORD* was still equipped with two lifeboats.

This happened because of misleading request from Brise Bereederungs and the flag state's failure to scrutinize all the information related to the safety of a vessel.

The flag state repeatedly approved the exemptions from the SOLAS regulations. MAIB analysis of exemptions issued by the flag state for the *CEMFJORD* in the 13 months prior to this accident found that the vessel was underway at 54 percent of this period with exemptions from safety regulations. The accident investigation report states that if the flag state had fully understood the situation of the *CEMFJORD*, it might well have refused to allow the vessel to depart from Rordal. These frequent safety exemptions came from global industry pressure on the flag state and RO.

The fact that the flag state did not scrutinize the requested content of safety exemptions and that they were willing to issue them without considering risk assessment of the vessel if the exemptions had been approved lead the *CEMFJORD* to deviate from SOLAS regulations.

Ineffectiveness of flag state inspections

The *CEMFJORD* had been inspected by Port State Control (PSC) inspectors and a flag state inspector who was employed by the Department of Merchant Shipping (DMS) Cyprus for seven years. The latter inspector conducted seven full inspections and seven documentary verification inspections, and sent the inspection reports with no deficiencies noted to

DMS Cyprus. During the same periods, PSC inspectors issued the extent of deficiencies.

There was no mechanism established in the flag state that alerted anyone to the different opinions between the PSC and the flag state inspector's reports.

Economic constraints

The *CEMFJORD* was operated by a charter agreement between the agent of owner and a charterer. The vessel was required to transport cement from Rordal, Denmark, to storage silos in various ports in Europe. The agreement prescribed the eight-hour of cement loading in Rordal and the navigating speed of about 9 knots, even though its maximum speed was 9.5 knots. The agreement did not allow recovering time for the *CEMFJORD* to arrive at planned destination if it had unexpected problems during loading and unloading or met with adverse weather conditions in the North Sea. In fact, the *CEMFJORD* started loading two hours later than scheduled, and departed the terminal five hours late on the day of the accident.

Tight scheduling came from commercial pressure and affected the master's decision-making, leading him to transit the Pentland Firth under severe sea and weather conditions.

4.6 Liveware-peripheral liveware

Central liveware-peripheral liveware interface represents the relationship between people. It includes communication, crew interaction, crew management and safety culture.

4.6.1 Communication

Lack of communication is often a contributing factor in accidents. Communication is composed of four components: Sender, receiver, message and medium (Parrott 2011). Face-to-face communication is the most effective, as senders and receivers can enhance their intended meanings with inflection, intonation and body language. When a sender and a receiver are located beyond mutual oral conversation range, they rely on devices such as radios, cellphones and sound-powered phones for communication. This communication mode requires them to speak with precision, use standard language, express clearly, listen carefully, confirm the message in turn and make sure that words do not overlap.

In order to ensure that the sent message is clearly understood by the receiver, a closed communication loop has been used onboard, for instance, giving steering order. The OOW is responsible for ensuring that the vessel is safely and efficiently maneuvered. When the OOW gives helm orders to the helmsman, the OOW needs to confirm that the orders were correctly followed in oral responses and actions taken by helmsman.

It is common that a ship's crew are multinational and use English as a common language and crewmembers can communicate each other. When entering ports and harbors, they face language barriers with pilots, tugboats, approaching local vessels and port authorities. Even though the master and pilots can communicate well, the master may not be able to understand the local language which is used between the pilots and tugboats. As a result, the master cannot comprehend the current and upcoming situation on the vessel maneuvering, impairing his ability to assume the situation, and leading to lack of situation awareness. The same thing happens when the master or the OOW encounter another approaching vessel and try to know its intended maneuvering to avoid close quarter situation, they often use VHF as a means of communication. Ironically, the fact that ship-to-ship VHF communication to avoid close quarter situations has been contributing to collisions, what is called "VHF assisted collision", should be noted. As mentioned above, VHF communication is not the most effective means to avoid a collision, rather, it gives the sender and the receiver a false situation awareness or enhances the lack of situation awareness, resulting in loss of an opportunity to take proper actions in ample time on both side of the vessels. These facts are illustrated in the accident report on a collision between the *KOTA DUTA* and the *TANYA KARPINSKAYA,* and VHF issues are discussed in the next section: "VHF assisted collision".

4.6.2 *VHF assisted collision*

Although the typical case of VHF assisted collision is described and illustrated in detail in the collision between the *KOTA DUTA* and the *TANYA KARPINSKAYA* in section 4.1, MAIB (2014b) also investigated a collision between the container vessel *MCA CGM Florida* and the bulk carrier *Chou Shan,* finding deficiencies using VHF as a collision avoidance action, and Marine and Coastguard Agency (MCA) issued warning on use of VHF.

Caution issued by MCA on use of VHF as a collision avoidance aid

In the wake of the MAIB accident investigation on the collision mentioned above, MCA circulated the Operational Guidance on the Use of VHF Radio as follows (MCA 2006).

(1) Using VHF radio in an attempt to avoid collision is not always helpful and may even prove to be dangerous.

(2) Agreement of manner of passing over VHF can disagree with the CORLEGs, and that may lead to danger of collision.

(3) Using VHF while vessels are approaching, valuable time can be lost to avoid collision in ample time.

(4) When they try to communicate in foggy weather, restricted visibility, or existence of multiple of vessels in the water, a sender and a receiver may not be identified. Furthermore, even if they are identified, there is still the possibility of a misunderstanding due to language difficulties which impair situation awareness.

(5) By the time both vessels realize that the ships are heading toward each other while following the agreement, the distance between them has reduced to the extent that neither vessel can prevent a collision.

4.6.3 Crew interaction

Crew interaction includes teamwork, crew and a pilot integration, and Bridge Resource Management (BRM).

A team is a group of people working together toward a common goal. The team has a leader, who is the most effective when filling both an appointed leader and a *de facto* leader. Also, each member of the team should know the role allocated to and the roles of other members. Consequently, a shared mental model is made up among the team, and the team members will understand how the work or vessel is going to complete it's purpose. The team members have the responsibility to tell the leader when they think the leader is making a mistake (Adams 2006, Swift 2000). The concept of teamwork can be applied not only to components of a bridge team but also to crewmembers in loading, unloading and other work related to maintenance.

The concept of teamwork and BRM is similar, but the applicable workplace is different. BRM is commonly used as a tool to manage the bridge team which, in general, consists of the master, the OOW, a lookout and a helmsman. The purpose is to reduce the possibility of chain-of-error, and have a decision-maker be able to make a correct decision to, for instance, avoid close quarter situations with another vessels. It originated from the cockpit resource management in aviation mode and did not intend to include an organization or ship management company, ship owner, and other organizations.

4.6.4 *Crew and pilot integration*

It is often commented that when a vessel starts to proceed with a pilotage, it is the beginning of a loss of situation awareness on the bridge team. Why does it occur? The pilots are familiar with the waters but relatively unfamiliar with the vessel, equipment and handling of the equipment and vessel. On the other hand, a bridge team that consists of the master and crew feel relaxed and expect the pilot to maneuver the vessel safely through crowded and rather unfamiliar waters. Language differences add to the communication breakdown at the bridge, especially when tug boats get involved in the pilotage. As a result, the bridge team tends to lose situation awareness (Adams 2006).

Parrot (2011) studied the number of accidents involving pilots in US waters and found that 90 percent of accidents occurred in confined waters, 60 percent of which occurred with a pilot onboard. He concludes that the arrival of the pilot onboard relates to loss of situation awareness shipside. He points out that, technically speaking, the pilot is an advisor to the master, and the master and crew hold the responsibility for the outcome of ship operation. However, the pilot, functionally speaking, directs the ship's maneuvering.

STCW Convention 1987 regulation II/I, "Navigation with pilot embarked" prescribes that the master and the OOW must co-operate closely with the pilot and maintain an accurate check of the ship's position and movement. Concerning the OOW, STCW Convention states that if the OOW is in any doubt as to the pilot's actions or intentions, the OOW should seek clarification from the pilot. If doubt still exists, the OOW should notify the master immediately and take whatever action is necessary before the master arrives.

Swift (2000) emphasizes that the ship's progress must be monitored by the OOW and reported to the master, and if any deviation from the planned track or speed is found, the master will be in a position to question the pilot's decision with diplomacy and confidence.

The next case distinguishes the issues of crew and pilot integration.

Product carrier *SICHEM MELBOURNE*

Summary: The *SICHEM MELBOURNE,* Singapore flagged 8,455 gross tons product carrier, departed a berth with a pilot onboard at the Coryton Oil Refinery on the River Thames estuary in the UK on 25 February 2008. While departing, the vessel made broadside contact with the jetty, sustained damage on her port side and damaged other mooring structures (Fig. 4.18) (MAIB 2008b).

The course of accident and contributing factors

The *SICHEM MELBOURNE* completed the discharge of gas oil at the Coryton Oil Refinery on 25 February, and planned to depart to the NE Spit anchorage at the entrance to the River Thames. At 19:51, a pilot boarded the vessel, and exchanged the information about particulars of the vessel with the master, but not about passage plans covering the departure operation.

The pilot advised the master that the last line to let go of would be the forward spring. The vessel was facing downstream, the wind was from the starboard quarter and an ebb tide was running. The pilot's intention of departure was to come ahead on the forward springs and let a wedge of water form between the ship's stern and jetty and then let go of the spring

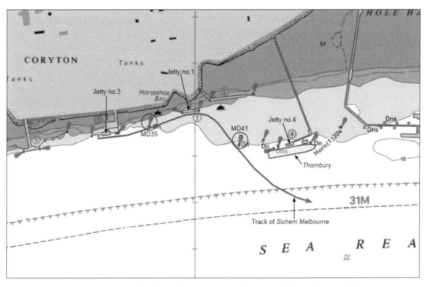

Fig. 4.18: Track of *SICHEM MELBOURNE* (MAIB 2008b).

lines and bring the vessel astern into the channel. The pilot thought that, considering environmental circumstances, his departure operation was self-explanatory and, therefore, did not share it with the master. On the other hand, the master's plan was, as the stream would be fairly slack due to an hour of ebb tide, to bring the vessel to go ahead with continuous port helm and bow thruster to starboard, proceeding transversely away from the berth on the course of 139°. If the stream was stronger, he believed tug assistance would be required.

During the departure operation, the pilot and the master were stationed on the port bridge wing. After the permission from the Port of London Vessel Traffic Services (PLVTS), the mooring ropes were singled up and then cast off, except for the forward springs. At 20:18, the pilot directed the helm hard to port and engines on dead slow ahead. The master ordered the bow thruster to starboard in order to prevent the ship's bow from pressing against the jetty. The pilot noticed the usage of bow thruster but made no comment about it.

Two minutes later, the pilot instructed stop engine as the angle between the ship's stern and jetty increased. The master assumed the pilot was ready to let go the forward springs and he ordered, in Russian, to let go and full bow thruster to starboard. The pilot who could not understand Russian noticed the ropes were released but still gave no comments to the master. The ship's port quarter started to move toward the jetty. The pilot thought he could recover the situation by moving ship's head to port and come astern into the channel. As the ship's bow thruster worked on full power to starboard by the master's intention, the ship's port quarter collided with the jetty No. 3. The vessel damaged jetty No. 3, and two other mooring structures until she reached the channel.

Even though there was no description on the investigation report about the pilot realizing that the master always took contradictory remedial maneuvers, the lack of communication about the departure operation and subsequent maneuver between the pilot and the master primarily contributed to this accident.

The official language on board *SICHEM MELBOURNE* was English, but the working language was Russian as the crew were of Ukrainian, Russian and Latvian nationalities. Communication made in Russian among crewmembers isolated the pilot from the bridge team.

The pilot had been employed by the Port of London Authority (PLA), which had a ship's bridge simulator, incorporating realistic simulation of the River Thames, but did not include any form of Bridge Team Management (BTM) training or master and pilot integration. The pilot had attended a BTM training at his previous company 13 years ago. There was no system established by the PLA on monitoring pilots' performance, such as level of pilot and master integration.

At the time of the accident, a southwesterly wind at Beaufort scale 4 (5.5–7.9 m/s) was blowing, and about 0.5 to 1 knot easterly setting tidal stream was going parallel to the jetty No. 3. According to the guidelines by the Petroplus of Coryton refinery, tugs were not required for ebb tide and head down unberthing. After the accident, Petroplus implemented a tug assisted maneuvering for all ebb tide head down departures from the jetties.

4.6.5 Crew management

Crew management includes supervision, task assignment, workload distribution and information dissemination from an organization. Vessel's operation and loading schedules, designation of the crewmembers, work/rest hours, role and responsibilities of all crewmembers, communication between ship and organization, etc., are all described in management rules, procedures which are in accord with flag state rules and regulations, which in turn are expected to incorporate IMO adopted conventions, codes and guidelines.

When considering crew management, there are some points that must be kept in mind: Group think and authority gradient. Hawkins (1987) explains that group think means that individual judgement may be influenced by judgement of the group. An individual can be influenced to make a wrong judgement if the majority of those around them have made the wrong judgement. The group decision may involve risk-taking.

On the other side of group think is authority gradient or power distance. The authority gradient was drawn to attention in the *Tenerife* disaster in 1977 when Pan Am and KLM Boeing 747 airplanes collided on a runway in foggy, i.e., restricted visibility, conditions at Tenerife-North Airport (formerly Los Rodeos). Five hundred and eighty-three people were killed in this disaster. The KLM pilots misunderstood the commands from the control tower, and started going down the runway to take-off where the Pan Am 747 was waiting for take-off. The KLM captain was one of the highest-ranking officers, and the copilot did not make an assertion to the captain's mistake. The authority relationship between the captain and the first officer/copilot in aviation has contributed to causes of many accidents and incidents. Hawkins (1987) shows that nearly 40 percent of first officers surveyed said that they had experienced failed communication of their doubts about the aircraft operation to the captain.

Optimum authority gradient is not too steep, like the KLM pilots, nor too flat. It allows them to have an effective interface among the pilots. Without it, errors made by the captain may go undetected and unrectified.

In marine mode, power distance and collectivism in container shipping was studied and it was found that it related to national culture (Lu et al. 2012). ATSB investigated and found that one of contributing factors of the grounding of the *BUNGA TERATAI SATU* was the authority gradient (ATSB 2001).

4.6.6 *Authority gradient*

Container vessel *BUNGA TERATAI SATU*

The *BUNGA TERATAI SATU*, Malaysia flagged 21,339 gross tons container vessel loaded with 875 containers, sailed from Singapore on 26 October in 2000 to Sydney Australia. The ship's crew consisted of three nationalities: Pakistani, Malaysian and Indonesian.

From 03:00 on 1 November to 05:00 the next day, the vessel proceeded via the inner route of the Great Barrier Reef with pilotage. After disembarkation of the pilots, the master took the conn until 06:35, then handed over the conn to the chief officer (C/O).

At 06:40, the duty able seaman (AB) stated cleaning the bridge while the C/O made a call on the mobile telephone at the bridge wing. The C/O invited his wife to the bridge and then made another call at the bridge wing to a relative's house in Karachi, Pakistan, where his children were cared for. He developed a practice of letting the AB plot the ship's position from GPS every hour in open waters.

The AB plotted the ship's position at 07:00 and found the position was adjacent to the waypoint on the passage plan. He assumed that the C/O would come back into the wheelhouse to alter the course. At 07:17, the C/O came back and went to the chart table to check the ship's position. He ordered to steer 180°. At that time the AB could see a sand cay on the starboard bow, and asked the C/O whether he wanted to turn to 180° towards the sand cay, adding that the ship was heading to shallow water. The C/O then responded turning the helm to 180° to port. The AB was confused and did nothing.

At 07:21, the *BUNGA TERATAI SATU* grounded on Sudbury Reef of the Great Barrier Reef at full sea speed of about 21 knots (Fig. 4.19).

Among other things that contributed to the grounding, ATSB stated that, after plotting the ship's position at 07:00, the AB noticed that the ship

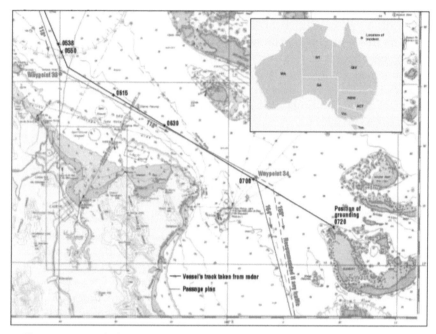

Fig. 4.19: Intended course and track of *BUNGA TERATAI SATU* (ATSB 2001).

had passed the waypoint and assumed that the C/O would direct him to alter the course in due time. He did not suggest the C/O to alter the course, instead resuming the lookout duties. Such an attitude reflects a power distance among the senior officers, junior officers and crew, and increases the possibility of chain-of-error.

ATSB also listed contributing factors as follows: (1) The C/O was preoccupied with arranging private telephone calls within coverage area of cellular phone while the ship was approaching a waypoint. The private telephone conversation distracted the C/O's attention from his duties as the OOW, i.e., monitoring the ship's position, course, speed, etc., (2) The ship's GPS cross-track error alarm was not audible enough to attract attention.

4.6.7 *Safety culture*

Safety culture has four subcomponents: Reporting culture, just culture, flexible culture and learning culture. These four subcomponents interact to create an informed culture, which is the equivalent of safety culture (Reason 1997). In reporting culture, people are prepared to report their errors and near-misses. In just culture, people are encouraged to provide safety-related information while definition is made between acceptable and unacceptable behavior. Under flexible culture, organizational structure shifts from hierarchical mode to a flatter professional structure during crisis. In learning culture, an organization draws lessons learned and implements reform if needed.

IMO (2003) defines the safety culture as "a culture in which there is considerable informed endeavor to reduce risks to the individual, ships and the maritime environment to a level there is as low as reasonably practicable."

There have been some accident investigation reports that investigated an organization related to the marine accident and analyzed the safety culture in the organization. Among them is the accident report on the investigation of the capsize and sinking of the cement carrier *CEMFJORD* issued by MAIB. The case is referred to in Chapter 10.

4.7 Conclusions

In this chapter, based on the SHEL model, the process of accident occurrence and its contributing factors are described in detail together with real accident cases and statistics on contributing factors. What is drawn from these accident cases and statistics is that the contributing factors of marine accidents do not necessarily fit all subclassifications of the five elements of the SHEL model, and that there are frequent contributing factors and relatively less frequent factors in the subclassification. It is also apparent that causal factors of marine accidents cannot be tracked down linearly, but the cause-effect link cannot be eliminated, convincing that marine accidents cannot be explained either by sequential accident model or systemic accident model. This chapter has reaffirmed that Late 1 SCM applies well to marine accidents. In other words, statistical analysis of contributing factors based on Late 1 SCM is considered to be able to provide readers an insight into the prediction of marine accident occurrence.

Chapter 5

Preparation for Accident Investigation

5.1 Introduction

Reason (1997) states that the process leading to an organizational accident proceeds in the order of an organization, a local workplace and an operator, but the process of accident investigation is the reverse: First an operator's unsafe act, then local workplace factors and organizational factors (Fig. 3.6).

In the real accident investigation, the processes proceed based on the evidence left behind at the accident site, and the process of interviewing an operator and witnesses on the transition of the events together with the collection of evidence is the most important part of on-site investigation. The accident investigators try to find an active failure that triggered the accident by interviewing operators, and looking at latent conditions, that were revealed by interviewing the management of their organization while collecting evidence at a local workplace and its organization, that support the content of interviewing, and by further investigating the environment that affected the occurrence of the accident.

Chauvin et al. (2013) and Schröder-Hinrichs et al. (2011) point out that such organizational factors as safety culture and organizational climate were not included in accident investigation reports issued by national marine investigation authorities. This is an issue that national investigation authorities across the world face today.

The accident investigation itself requires investigators to comprehend human errors, risk management at workplaces, ISM code at organizations, and utilize them for accident analysis among other things. This chapter mainly focuses on human error, risk management and ISM Code as a facilitation to extensively find out both local workplace and organizational factors.

5.2 The human error

The human error is closely related to the Human Factors described in Chapters 2 and 4. It is defined as an action or non-action that did not achieve a desired goal, as a result of human capabilities and limitations. In this section, the historical transition of human error theories and classification of human error are described briefly.

5.2.1 Transition of human error theories

The human error theories transitioned from Freud's theory through Norman's theory and Rasmussen's theory to Reason's theory. These are summaries of each theory (Strauch 2004).

Freud's theory

Sigmund Freud stresses that error is a product of the unconscious drives of the person. Those who made errors are less effective and more deficient than those who did not. As a result, certain people are more likely to make errors than others due to their particular traits of those people.

Norman's theory

Donald A. Norman (1988) emphasizes that people commit some errors in the error-prone settings which are unrelated to particular traits of those people. He differentiated between slips and mistakes: Slips are action errors, and mistakes are errors of thoughts.

Rasmussen's theory

Jen Rasmussen expanded Norman's theory by defining three types of performance and errors: Skill-based, rule-based and knowledge-based. He states that people perform their actions at these three different levels. Skill-based level is used when they carry out routine, highly practiced tasks in an automatic manner. When it comes to rule-based level, they use stored rules such as if these situations are encountered then do these actions. Knowledge-based level is the highest level of performance, such as in a novel situation, people are forced to work very slowly using all resources stored in the memory with trial and error learning. He stresses that when people make decisions, the process is not linear, they often shortcut processes in real life situations. As a result, these three types of level shown above can coexist at any time (Rasmussen et al. 1987).

Reason's theory

James Reason enlarged Rasmussen's theory. He states that slips and lapses are action errors that involve skill-based performance. Mistakes and violations are errors of people's intention, and mistakes occur either in rule-based or knowledge-based performance. He developed latent conditions (previously called latent errors) and differentiates between active failures (previously called active errors) and latent conditions: Active failures are committed by operators, and latent conditions which lie dormant for a long period are committed by higher levels, such as designers and managers. Active failures are influenced by latent conditions, and accidents occur due to a rare combination of these errors (Reason 1990).

His theory became the source of development of the SCM in 1990, the model which can illustrate human contribution to a complex system failure and later influenced ICAO and IMO in terms of facilitating the understanding of human factors issues and safety on the investigations.

5.2.2 Classification of human error

In general, an accident is triggered by an active failure; investigators collect comprehensive evidence on human factors, local workplace factors and organizational factors starting from the active failure based on the concept that an incident and accident occur due to a failure of the system not simply to a failure of an operator. An error is a failure of planned action to achieve a desired goal, and human error is a natural part of life. Everybody makes errors. However, the active failure in the SCM is defined as an unsafe act which is committed in the presence of a hazard or in a hazardous situation. If an error is committed outside the hazardous situation, it will not trigger an incident or accident.

IMO (2014) adopts the human error classification based on the Cognitive Reliability and Error Analysis Method (CREAM) developed by Hollnagel (1998) and Generic Error Modeling System (GEMS) by Reason

(1990). CREAM divides human error into action, observation, planning and intention and interpretation. GEMS divides an unsafe act into an unintended action which includes slips and lapses and an intended action which includes mistakes and violations. Slips and lapses are execution failure (e.g., a pilot pushed a wrong button for sounding signals unintentionally towards an approaching vessel) and storage failure (e.g., the master thought their own vessel had passed a light buoy which was a waypoint and then altered the ship's course, whereas, in reality the vessel had not yet passed the light buoy) in the cognitive stage, respectively. Mistakes occur during the planning stage (e.g., the master drew the short-cut course in a channel on chart, resulting in head-on collision situation with a meeting vessel). Slips, lapses and mistakes are designated as basic error types.

5.3 Risk management

5.3.1 Risk management process

As mentioned in Chapter 2, understanding risk management processes, quality management systems and ISM Code in the shipping industry is needed in order to find a situation surrounding a vessel, an operator, and an organization before and after the accident.

According to IEC/ISO 31010 (2009), the risk management process involves risk identification, risk analysis, risk evaluation, risk treatment and monitoring and review (Fig. 8.2 in Chapter 8). Risk identification means identifying the causes and source of the risk, event, situation, circumstances and the nature of that impact. Risk analysis is to determine a level of risk by measuring the consequences and probabilities for identified risk events. There are three methods for analysing the risks: Qualitative, semi-quantitative and quantitative. Among these risk analyses, semi-quantitative and qualitative are commonly used because quantitative analysis is not desirable due to there being insufficient information about the system or events being analysed, lack of data, and implication of human factors. Risk treatment is to select and implement options that could change the level of risk, that is consequences or probabilities for risk events or both. Monitoring and review mean that risks and control should be monitored and reviewed periodically.

In actual cases, COLREGs, passage planning, work procedures such as enclosed space entry incorporate the risk management process. For instance, the cargo vessel *LANA* ran aground on Yokone, about 6 m height rock, located between the Izu peninsula of mainland Japan and Mikomoto Shima, the small island off the cost of the Izu peninsula. The master selected the intended route, which passes between the Izu peninsula and Mikomoto Shima, during the planning stage of passage planning, before departure from Chiba port in Japan. Sailing directions published by the Admiralty in the UK recommend ocean-going vessels should not travel the channel but proceed via the southern water of Mikomoto Shima because vessels will encounter strong current and dangers like rocks in the channel. The master studied the sailing directions, knowing the existence of danger in the channel, and decided to pass via the channel during the night. He had experienced passing through the channel several times during the daytime. The intended course was drawn on the chart by the master, and, using GPS, the ship's progress was monitored occasionally by an able seaman (AB) who was not required to learn fixing ship's position in STCW Convention. The master did not monitor the ship's progress by himself and assumed the ship's plotted positions made by the AB on the chart would be correct. The master kept an eye on the distance and bearing from Mikomoto Shima. The master did not notice the deviation from the ship's intended course until the *LANA* ran aground on the rock. It was found that risk analysis and monitoring and review processes of the risk management process were not properly conducted by the master (JTSB 2011c, Fukuoka 2016a).

5.3.2 *Methods of risk analysis and evaluation*

In practice, most mariners determine the risk level based on their past experiences and norms of shipping industry. In cases of occupational casualties, mariners or workers have been facing various kinds of risk events and levels of risks. In such cases, past experiences are not always applied because a number of accidents have occurred due to lack of or insufficient risk management, so detailed risk analysis and evaluation are needed. There are established methods for analysing and evaluating risks, the ones that follow are part of the methods that could be useful for the shipping industry, especially when addressing the number of incidents and accidents.

Risk analysis employs severity scales and frequency scales. There are a number of types of scales, the nominal scale label and ordinal description. Maguire (2006) provides the scales shown below that are very clear and easy to measure. Table 5.1 is adapted from the UK defence industry. Maguire adds that Table 5.1 illustrates the relationship that one death is equal to 10 severe injuries and that this ratio is open for ethical debates. Table 5.2 shows the frequency scales, leaving some ambiguity about probable and occasional. Table 5.3 provides more clear definition on the frequency scale, but it might not be suitable for some industries that require more detailed scales.

Risk evaluation uses a risk matrix which is a combination of severity scales and frequency scales. The risk matrix shows whether an identified risk event can be tolerable or acceptable by an organization a reader belongs to, and this decision-making should be communicated and agreed upon in the organization. Safety requirements or demands surrounding the organizations have been changing from lenient to stricter ones, as seen in the shipping industry. In other words, criteria of whether something is acceptable or not to the organizations will be ever changing over time.

According to IEC/ISO 31010 (2009), the risk level is divided into four categories.

(1) The risk cannot be justified except for in extraordinary circumstances.

(2) The risk can be tolerable only if benefit received grossly exceeds the risk.

(3) The risk can be tolerable if benefit received exceeds the risk.

(4) The risk is broadly acceptable.

Unacceptable level of risk is (1), ALARP (As low as reasonably practicable), (2) and (3) are tolerable, and (4) is broadly acceptable. The risk matrix employs this concept, called ALARP concept, labeling from A to D in order of risk level. Table 5.4 illustrates a sample risk matrix (Maguire 2006).

Table 5.1: Severity scales.

Nominal scale	Ordinal description
Cataclysm	Thousands of fatalities and injuries
Disaster	Hundreds of fatalities and injuries
Calamity	Tens of fatalities
Catastrophic	Single figure, multiple deaths (up to 10)
Critical	Single death and/or multiple severe injuries
Severe	Single or several serious injuries
Significant	Several minor injuries-significant loss time
Minor	Single minor injury

Table 5.2: Frequency scales (1).

Nominal scale	Ordinal description
Frequent	Likely to be continually experienced
Probable	Likely to occur often
Occasional	Likely to occur several times
Remote	Likely to occur some time
Improbable	Unlikely, but may exceptionally occur
Incredible	Extremely unlikely

Table 5.3: Frequency scales (2).

Nominal scale	Ordinal description
Frequent	Weekly
Probable	Monthly
Occasional	Annually
Remote	10 yearly
Improbable	100 yearly

Table 5.4: The risk matrix (Maguire 2006).

	Calamity	Catastrophic	Critical	Major	Minor
Frequent	A	A	B	C	C
Probable	A	B	B	C	D
Occasional	B	C	C	D	D
Remote	C	D	D	D	D
Improbable	D	D	D	D	D

Before starting work, operators and organizations are required to make sure that the risks are within tolerable or broadly acceptable regions. If the risks remain, they need to be reduced in accordance with priority ranking in risk reduction methods.

For instance, prior to entering an enclosed space, such as a cargo hold or pump room, operators must measure the level of atmosphere at different heights above the floor level inside the space using an oxygen analyzer or other devices. If the level of oxygen concentration is less than 21 percent, the risk is at an unacceptable level, they must forcibly ventilate the space to raise the level of oxygen concentration. Even if the oxygen concentration has been improved to 21 percent, while working in the space, operators must periodically measure the oxygen concentration in order to ensure that the atmosphere inside the space remains safe, which is the process of monitoring and review. If the level of concentration deteriorates during work, operators are required to interrupt the work, retreat to a safe area, and restart forced ventilation.

5.4 ISM Code

5.4.1 Background of IMS Code

Safety management system and ISM Code are developed by ISO series, part of which is the quality management system, and are the productions of history of self-regulation. This section first looks at the history of ISO series and ISM Code and then transition to self-regulation. The numbers following the titles of the ISO publications denotes their year of publication.

(1) ISM Code was adopted by IMO Resolution A.741(18) on 4 November 1993, based on ISO 9002.

(2) ISO revised ISO 9000 series which included ISO 9000, 9001, 9002, and 9004 in the 1990s.

(3) ISO 9000:2000, ISO 9001:2000 and ISO 9004:2000 were published in 2000 after integrating ISO 8402:1994, ISO 9000-1:1994, ISO 9001:1994, ISO 9002:1994, ISO 9003:1994 and ISO 9004-1:1994.

(4) ISO 9001:2000 integrates ISO 9001:1994, ISO 9002:1994 and ISO 9003:1994 and only stipulates requirements of the Quality Management System.

(5) ISO 9004:2000 states that an organization can identify, assess and manage the risks related to interested parties by monitoring the ever-changing organization's environment. ISO 9004:2000 provides a wider focus on quality management than ISO 9001:2000, however, it does not cover risk management, instead, it states that more information on risk management be referred to ISO 31000. ISO 31000:2009 provides principles and generic guidelines on risk management.

(6) ISM Code was revised by MSC.273(85) on 4 December 2008, based on ISO 9001, not ISO 9004. This means that ISM Code does not incorporate the guidance of ISO 9004 which provides organizations with assistance for the achievement of sustained success under ever-changing environments by a quality management approach.

5.4.2 History of self-regulation

In the UK, legislative bodies had established legislation on health and safety in detail, and the regulatory body had made individuals and organizations comply with these enacted rules by legal grounds in the first half of 20th century. The safety legislation was characterized by fragmentation and lack of workforce involvement. It imposed duties on employers. As the industrial accidents rose sharply in the 1960s, public desire for legislation concerning health and safety increased, a Commission of Inquiry was established under Lord Robens, who was chairman of the National Coal Board. The Robens Report was submitted to the UK government in 1972. In 1974, the Health and Safety at Work Act reflecting the report was enacted. After this law was established, the ideology of self-regulation on health and safety rules have increased since the 1990s, and the environment changed to voluntary regulations that organizations autonomously create safety self-regulation and decide on their own. After that, not only in the UK, but also in many other industrialized countries, the way of establishment of safety-regulation has changed greatly. The law has shifted from something to designate means to achieve work safely to something that tries to achieve a specific safety goal (Reason 1997).

In the shipping industry, the voluntary ISM code came into effect for oceangoing vessels from 1998, and in 2002, it was applied to all oceangoing vessels. Also, in some countries, non-oceangoing vessels also adopt the ISM code voluntarily. Under these circumstances, the organization's safety culture is now playing a greater role in organizing and continuously maintaining self-regulation.

5.5 Core of quality management system

The core of the quality management system is represented by the PDCA cycle (Fig. 8.3 in Chapter 8). For continuous improvement, the PDCA cycle needs to be reiterated. Iizuka (2009) describes the effective PDCA cycle that an organization should establish and operate as follows.

Plan

(1) Clarify target to be managed. For example, in the shipping industry, reduce the number of marine incidents and accidents of vessels under an organization's management from the previous year.

(2) Determine the scale to achieve the purpose. In the case of the above sample, it becomes percentage, or the number of vessels involved in incidents and accidents.

(3) Determine the level or target that management items should achieve. For instance, reduce the number of marine accidents by 10 percent every year, and eliminate it after 10 years. In the manufacturing industry, management items are selected by utilizing the Pareto chart, which illustrates "vital few, trivial many." The concept was derived from the fact that the total population of 20 percent held 80 percent of the total land in the nation. Quality management applied the concept to problems of product quality: 80 percent of quality problems attributed to 20 percent of causes. This implies organizations must choose important items from a statistical standpoint.

(4) Clarify ways to achieve these goals. The ways mean various kinds of operation procedures.

Do

(1) Conduct education and training on operation procedures for operators to carry out work. For instance, prior to entering enclosed spaces, operators are required to learn and practice measuring the atmosphere inside the spaces with devices, responding to cope with deteriorated atmosphere, taking emergency measures if an accident occurs.

(2) Do the work according to operation procedures. If a good result, that the procedures should produce after being followed correctly, is not achieved, operators need to offer the organization deficiencies of operation procedures. The organization revises the operation procedures by verifying the deficiencies that have been offered. By doing this, it is possible to prevent the operation procedures and the actual works operated at workplaces from being mismatched.

Check

In light of the goal, check the process and results based on facts and find out whether there is a problem. Internal and external audit, management review, accident investigations and walk-through correspond to this process. Accident investigations should be conducted with scientific methods.

Act

During the Check process, if there is a deviation from the goal, measures are taken. There are three measures: Correction is taken to eliminate a detected nonconformity, and corrective action is taken to prevent recurrence. Preventive action is taken to prevent occurrence by identifying problems during the Do process and incorporating measures to be taken against the problems beforehand.

5.6 Contents of ISM Code

These are part of provisions which prescribe the core of ISM Code. The numbers in the square brackets denote the clause of the ISM Code.

Plan

(1) The objectives of the ISM Code are to ensure safety at sea, prevention of human injury or loss of life, and avoidance of damage to the environment and to property [1.2.1].

(2) The objectives of safety management of the organization should: (a) provide for safe practices in ship operation and safe working environment, (b) assess all identified risks to its ships, personnel and the environment and establish appropriate safeguards, (c) continuously improve safety management skills of personnel ashore and aboard ships, including preparing for emergencies [1.2.2].

(3) The safety management system should ensure: (a) compliance with mandatory rules and regulations, (b) that applicable codes, guidelines and standards recommended by the organization, administrations, classification societies and marine industry are considered [1.2.3].

(4) The organization should: (a) establish a safety and environmental protection policy which describes how the objectives will be achieved, (b) ensure that the policy is implemented and maintained at all levels of organization and workplaces [2.1 and 2.2].

Do

(1) The organization should establish and maintain procedures for identifying any training which may be required in support of the SMS and ensure that such training is provided for all personnel concerned [6.5].

Check

(1) The SMS should include procedures ensuring that non-conformities, accidents and hazardous situations are reported to the organization,

investigated and analyzed with the objective of improving safety and pollution prevention [9.1].

(2) The organization should carry out internal safety audits onboard and ashore at intervals not exceeding twelve months to verify whether safety and pollution prevention activities comply with the SMS [12.1].

Act

(1) The organization should establish procedures for the implementation of corrective action, including measures intended to prevent recurrence [9.2].

5.7 Implication of the PDCA cycle in an accident

The fire case implies the importance of improving safety using the PDCA cycle. This section, while describing an event chain, focuses on each process of the PDCA cycle surrounding the education and training on fire drill because the chief engineer (C/E) died from suffocation due to carbon dioxide intoxication (JTSB 2011d).

Fire of the car carrier *PYXIS*

Summary: On the *PYXIS*, 43,425 gross tons Panama flag car carrier, loaded with 3,900 cars, while proceeding in the high seas off the coast of Kinkazan, northeastern Japan, a fire broke out in the car deck at about 09:48 on 14 October 2008. Although the fire was extinguished by CO_2 (carbon dioxide) gas released into the car decks from fixed fire-extinguishing systems, the C/E died while evacuating the engine room. The accident occurred because the fire broke out in the engine bay of a car carried on the portside of Deck 10, and the fire spread to the cars nearby and on the upper decks. Despite detailed investigations, including experiments and simulation, the cause of the fire in the engine bay of the car remained unclear.

The course of accident

At 09:48 on 14 October, a fire warning on the control panel installed at the bridge went off and an alarm followed. The 3/O, OOW on the bridge watchkeeping, immediately reported the occurrence of the fire to the master. The master rushed into the bridge, confirmed the location of the fire with the fire detection system and instructed the 3/O to identify the site of the fire. The 3/O rushed to Deck 10, opened the fire door at the entrance, saw a bright yellow light and reported to the master. The master instructed the 3/O to return to the bridge. The 3/O, on the way to the bridge, checked Deck 11 and 13, found the smoke spreading and filling through these decks.

The master decided to use the fixed CO_2 fire extinguisher system to put out the fire, and instructed the C/O to close the dampers. He also instructed the 3/O to sound the general alarm and make the PA announcement of the fire-breakout in order to notify the deck department crewmembers who were checking the lashing status of the cars on the car decks and the engine department crewmembers working on maintenance in the engine room. The 3/O sounded the general alarm at 09:53 and repeatedly made announcements on the PA of "Fire, fire, fire, fire on Deck 11." A/Bs went down the stairs in the engine room to Deck 6 to inform the deck department crewmembers of the fire, and went up to the muster station on the boat deck.

In the engine room, the C/E, the first engineer (1/E) and other members, upon hearing the general alarm, went to the engine control room and confirmed by a call to the bridge that the location of the fire-breakout was Deck 11, cargo hold 3.

The 3/O, after having shut down the breaker for the lighting power line on the car deck, was instructed by the C/O who had received the instruction from the master to announce that all crew should go to the muster station and be ready for the release of CO_2 at 10:04.

The C/E and the 1/E, who had stayed in the engine control room because that was the room designated as the muster station in the case of a fire, heard the PA announcement that all crew should go to the boat deck, which was the muster station for abandoning ship. The 1/E proposed the C/E to go up to the boat deck together, but the C/E replied that he would stay here and be the last person to arrive at the master station. The 1/E went up by himself onto the boat deck.

At 10:08 the master confirmed with the 3/O whether all the crew were assembled at the muster station and found out that the C/E was not there. At 10:10, the master, in the fixed CO_2 fire extinguisher system room at the stern end of the accommodation space, received from the boatswain that all the dampers were closed, and started the release of CO_2 into Zone F, where the smoke sampling type fire detection system had detected smoke. The master expected that the C/E was in the engine control room away from Zone F.

At 10:14 the master received the report that the smoke sampling type fire detection system had sounded alarms for zones E and D, thought that the fire had spread to those zones, and released CO_2 into the zones. The master, in order to cool the car deck, instructed to discharge water onto the boat deck, around the position corresponding to the ceiling of Deck 13, and released CO_2 again into zones F, D and E at 11:07 because the master was not certain that the fire had been extinguished. At 11:16 all the CO_2 had been exhausted.

The *PYXIS* made a full turn in order to return to Japan, and at 07:00 on 15 October, met a Japan Coast Guard (JCG) patrol boat that had come to rescue it at the request of the ship management company. At 07:58, the master accepted six members of the JCG special rescue team on board in order to investigate the situation of the fire and the search and rescue of the C/E. At 10:20, JCG confirmed that the fire was extinguished, and at 10:45 it found the C/E's body near the car ladder on the starboard quarter end of Deck 7, trapped between a car and the bulkhead of the engine room.

Responses to fire prescribed on the Safety Management System

Fire-fighting procedures written on the SMS onboard the *PYXIS* were as follows: (1) when a fire is detected, crew raise the alarm and then fight the fire after help arrives, (2) the crew report to the OOW, (3) crew around the fire take initial firefighting activity until firefighting group arrive the scene, (4) the OOW contacts the master and order firefighting station, (5) engineer on duty breaks down electric source connected to fire location, (6) the OOW operates CO_2 fire extinguishing system only by master's command.

Fire drill conducted by *PYXIS* before the accident

Education and training on emergencies including fire and abandoning ship was conducted monthly onboard the *PYXIS*, according to SOLAS Convention. Content of the fire drill done on 11 May 2008 was as follows.

(1) At 14:25, assumed that fire was discovered at galley's hot plate. Chief cook applied initial fire-fighting using a portable CO_2 extinguisher.

(2) Motor man shouted "Fire at the galley" repeatedly and ran to the bridge to inform the OOW.

(3) The OOW immediately called the master and reported the fire station.

(4) The master ordered the OOW to raise the alarm, and followed it up with a PA announcement.

(5) The crew mustered to their respective stations.

(6) The C/O confirmed that there was the fire at the galley, closed all ventilation, fire doors and shut off electrical power, laid down fire hoses and ordered the engine room to start emergency fire pump.

(7) The fire back-up group applied boundary cooling.

(8) The C/O acknowledged that the fire was uncontrollable, and immediately reported to the bridge.

(9) The master ordered crew to back-off and prepare to abandon ship.

Evaluation from viewpoint of PDCA cycle

Comparing the core of quality management system or the PDCA cycle with the accident occurrence situation, content of SMS and drill on fire carried out onboard, it becomes clear where the defect was. The items listed below describe deviation from each process for coping with the fire that occurred in the car carrier *PYXIS*.

(1) Plan process was insufficient. The ship management company established fire-fighting procedures whose content was too ambiguous to be practical in the case of a real fire. For instance, it was not clear on procedures to contain a fire by blocking oxygen supply, steps to use CO_2 fire extinguishing system, actions when some crew did not appear at the muster station, clear and safe evacuation route and assembly point when using CO_2 fire extinguishing system.

(2) Do process was not conducted. Education and training on CO_2 fire extinguishing system to all crewmembers was not carried out. A very general training on coping with a fire in the galley was conducted, but training on cargo fire or automobile fire which was particular to the car carrier had not been conducted.

(3) Findings of internal or external audit were not written in the marine accident report, so Check and Act processes cannot be evaluated. However, the investigation and its findings themselves by national marine accident investigation authority are partly regarded as Check process, and the ship management company needs to reflect the findings written in the accident report to take correction and corrective action during the Act process in order to enhance safety.

5.8 Conclusions

The accident investigation must cover local workplace factors and organizational factors, but it is pointed out that the actual accident investigation reports are lacking in the investigation into and analysis of the organization factors. As seen in detail in Chapter 4, these factors are often caused by deviations between planning and executions. The local workplace factors can be revealed by making use of the concept of the risk management process, and organizational factors are clarified by using the theory of the PDCA cycle, comparing each process with the situation of the local workplace or vessel and the organization at the time of the accident.

Chapter 6

On-site Investigation

6.1 Introduction

The investigation process consists of data collection, analysis and report writing. Insufficient data collection affects the quality of the investigation report. ICAO (2001) states that the outcome of an accident investigation depends on the skill and experience of the investigators assigned to it, and that investigators need not only technical skill, but also integrity and impartiality in collected data, logic and perseverance in pursuing contributing factors, and ability to relate well with people who were involved in the accident.

This chapter discusses the safety of investigators on accident sites, protection and collection of evidence, and assessment of the evidence.

6.2 Health and safety

6.2.1 General health and safety

The site of the accident investigation is immediately after the accident, and there are various levels and kinds of risks. Although safety is everyone's responsibility, investigating organizations have a duty to establish effective procedures, good management on safety and proper education and training to investigators. Investigators need to ensure their own health and safety by assessing the risks they will encounter before entering a scene and by continuing risk assessments on-site during the course of investigation. Principles of safety on-site are as follows: (1) all foreseeable hazards need to be identified, (2) unacceptable risks need to be eliminated entirely, and (3) during the step of risk reduction, risks need to be reduced to a tolerable level or ALARP.

A generic or static risk assessment will be conducted prior to the on-site investigation by an investigation organization, and dynamic risk assessment needs to be conducted by investigators at the accident scene during on-site investigation. Static risk assessment is performed based on such available information prior to on-site investigation as notification of accident occurrence, information from an operator and a ship management company, cargo lists, location of accident occurrence, and advice from experts in dangerous goods onboard. Dynamic risk assessment is based on hazard identified on-site, detailed information obtained from coast guard and a fire-brigade, medical information from doctors and medical personnel, and analysis from experts.

The last resort in risk reduction methods is donning proper PPE. Based on the information obtained before an on-site investigation, the level of PPE is estimated, but as a result of observing the hull and cargo holds, loaded cargoes, etc., on accident scene, investigators might notice the difference from the hazard obtained beforehand. In such a case, it is also necessary to temporarily suspend the on-site investigation and reduce the risk or to wait until the risk is reduced to tolerable level, instead of wearing a higher-level PPE which might additionally require a plastic coated suit and a full face respirator with filter cartridge.

6.2.2 Hazards to consider

When accidents occur in extremely cold or hot seasons or regions, weather conditions such as temperature, humidity, snow and wind speed become a hazard, and then measures against cold or hot conditions must be taken in order to prevent frostbite, heat stroke, dehydration and so on. Extraordinary weather phenomena, such as flash flooding, landslides and storm surges in the wake of hurricanes and typhoons need to be monitored by gaining accurate and updated weather information from reliable authorities.

Prior to an on-site investigation, it is necessary to obtain a list of hazardous substances from the ship management company or an owner and to grasp the place of their installation because there are cases where explosives, radioactive substances, and toxic substances are loaded on the ship.

In the working environment, a ladder attached to a small vessel which investigators are boarding may not be properly equipped, for instance, the ladder is bent or missing some rungs. Investigators often use a pilot ladder to embark an anchoring vessel. Concerning safety of a pilot ladder, Japan Federation of Pilots' Associations (2012) reported that there were 31 accidents, including nine deaths of pilots boarding and disembarking from 2002 to 2011. The most frequent accidents were fall from a pilot ladder because pilot ladders were not completely secured to the hull. The pilot ladder is required to be properly equipped in accordance with SOLAS Convention.

Special attention to explosion-proof is required when boarding a tanker. Explosion proof clothing and shoes are required and taking photos are not allowed in most places. Investigators need to ask a responsible ship officer for permission.

Investigators have lots of opportunities to visit a shipyard by confirming the damaged part of the vessel after an accident. It is necessary to keep in mind that the shipyard is filled with hazard, for instance, fire work on deck, work tools and materials scattered on deck, multiple works going on with plenty of workers, moving objects like a crane, unsecured cargo hold and hatches, and noises that make it difficult to hear communications clearly. Thus, investigators need to consider when and how to conduct investigations at the shipyard safely.

When investigating a vessel which sank following an accident and that is planned to be floated by salvage works, it is necessary to be aware of the following points. The salvage program is subject to change due to weather and progress of floating works, investigators need close communication with a person in charge of salvage work. When conducting reconstruction of accident with witness or crew, it is essential to ensure the available time and safety of all persons on board. Entry into compartments or engine room and other structures below the upper deck ensues risk; expert advice is required prior to commencing. When entering the engine room, poor lighting, oily and slippery floors and ladders can be hazardous situations.

When entering an enclosed space, ensure all such hazardous situations, such as oxygen deficiency, toxic gases, nitrogen, hydrogen sulfide and carbon monoxide are avoided. Germanischer Lloyd (2011) sets up prerequisites for tank entry. First, permit for entry is issued by the master. Second, a responsible ship's officer is ready. Third, gas free certificate is provided by a class. Fourth, means of communication are arranged. Finally, safety and rescue equipment are positioned.

In the case of a fire accident, investigators will conduct an accident investigation after the fire fighting activity, thus, the following things concerning composite materials and fiberglass are considered (ICAO

2001). As fiber dust is hazardous to the eyes, skin and respiratory system, investigators need to remain upwind and wear disposable coveralls, goggles and face masks. Splinters from fractured fiberglass and composites should be handled with gloves.

In occupational casualties, investigators often encounter blood or body fluid from dead bodies on-site. Particles of dried blood may become airborne and come into contact with human mucus membranes. Preventing those from entering the eyes, nose, mouth and open lesions is essential. Since it is not possible to confirm that the blood or body fluids left on the scene are infected, accident investigators need to recognize that they are constantly exposed to biohazards. As a precaution, investigators are recommended to have inoculation against hepatitis B and other infectious diseases.

In some countries, when conducting an investigation, some kinds of inoculations are required to have a number of doses for effectiveness. IMO (2014b) recommends the inoculation of Yellow fever, Tetanus, Diphtheria, Polio, Hepatitis A, Hepatitis B and Typhoid depending on the location of on-site investigation, while considering investigation schedule.

In addition to these, it is necessary to obtain accurate information about local security situations, outbreaks of infectious diseases, food poisoning and other issues related to hygiene.

6.2.3 PPE and blood contamination

ICAO (2001) recommends the following as preventive measures on-site safety when encountering blood or body fluid.

(1) Areas contaminated by blood or body fluids should be made clear by rope and sign showing single point of entry and exit.

(2) Only persons properly equipped with PPE should be allowed inside the area. While donning PPE, eating, drinking, touching face, eyes, mouth and nose are prohibited.

(3) Humid and hot weather can result in heat stroke, consequently, it is important to manage the amount of time working with PPE.

(4) Used PPE must be disposed of.

(5) Exposed skin should be wiped with moist towel and washed with soap and water or disinfection chemicals.

(6) Proper PPEs against biological hazards, including blood and body fluid, are disposable latex gloves worn under work gloves, face masks

which cover the nose and mouth, protective goggles fitted with one-way check valves or vents, disposable protective suits that are durable and liquid-resistant, disposable shoe covers made of polyvinyl chloride or butyl rubber and protective boots made of leather or rubber. In addition, disinfection chemicals with rubbing alcohol of 70 percent strength and biological hazard disposal bags colored red or orange and labelled as Biological hazard are required.

In addition to these recommendations, in order to protect investigators from harm, PPE must be properly donned and disposed of. These are points to be noted on donning and undressing PPE.

(1) When donning PPE, it is necessary to take a buddy system and confirm with each other whether the PPE is correctly worn or damaged.

(2) The goggle may become cloudy because the air leaks from the mask between the nose and the eyes. At that time, curve the mask so as to be in close contact with the face.

(3) When undressing a protective clothing, first, the head hood is wrapped from the inside to the outside, then the torso and the leg are sequentially turned from the top to the bottom.

(4) Disposable gloves are usually worn with two different gloves with different colors. When one gloves is damaged, it is noticed by the difference in color, and the second glove can protect against blood contamination. The gloves are removed from the wrist part and wrapped from the inside to the outside.

(5) When it is necessary to undress PPE, such as at the completion of the on-site investigation or temporary interruption of investigation, investigators must undress PPE with prescribed undressing method at the designated place and discard disposable gloves and masks into designated garbage bags in order to prevent cross contamination.

(6) After decontamination, investigators undress protective clothing and take a rest at the designated place. Make sure that people wearing protective clothing do not mix with people undressing during the break.

6.2.4 Risk assessment sheet

The risk assessment sheet can help investigators to assess risks during on-site investigation. The sheet includes categories of risk, types of hazards, situation of hazard, location that the hazard identified, level of risk, its priority number, and risk control. The names of combustibles, structures, explosives, etc., are listed in the category column, and in the case of combustibles, estimated hazards, such as fuel and fuel tank, flammable liquid and oxygen leakage, are filled in type of hazards, and their condition and locations onboard the ship are entered in the column. Level of risk is put into the column, for instance, referring to the risk matrix shown in Chapter 5. Risk control is conducted in accordance with the priority ranking of risk reduction methods shown in Chapter 2.

Even when new hazards are identified on site, investigators must follow the same process as above and constantly monitor the risks and ensure the safety of investigators and people during on-site investigation.

6.2.5 Cordon

There are two types of cordon: Inner cordon and outer cordon. Generally, when an accident occurs, an inner cordon which surrounds main damaged bodies, such as the hull of a ship, fuselage of airplane, compartments of train and a scattered debris, is set up. Outer cordon is established on the outside of it in order to regulate rescue work, evidence collection and entry of outsiders. In marine accidents, unlike an airplane and a railway accident, since on-site investigation is often completed in the vessel itself, it is not common to have both types of cordons. However, when if a cruise ship collided with the facility of a pier at a tourist spot and a large number of casualties occurred, two cordons would be needed in order to ensure safety of people such as tourists, preservation of evidence, control of on-site investigation, coordination with media and so on. In some countries, legal systems where the criminal investigation takes precedence over the safety investigation allow judicial authorities to have responsibility for setting these cordons.

6.3 Evidence

6.3.1 Evidence in general

Evidence can determine the sequence of event, support robust analysis and justify recommended safety issues. Evidence is divided into four categories: human evidence; physical evidence; documentary evidence; electronic evidence.

Principle of collecting evidence is to collect perishable evident first, and then the rest. Perishable evidence includes human evidence, electronic evidence such as VDRs, AIS, GPS, VTS recording, CCTV, switch positions of radar and other navigational equipment, physical evidence such as pre-rescue scene, metallic fracture surfaces, ground marks, fuel and oil spills, atmosphere in cargo tanks and enclosed spaces.

General approach for collecting evidence is as follows (IMO 2014b):

- Interviewing witness
- Measurement, including sketching and mapping of the position of debris, equipment, injured persons or persons related to the accident
- Photography
- Examination of documentation, including charts, passage planning, standing orders, log books, cargo records, loading plan, SMS manuals, education and training records, internal and external audit records and statutory instruments
- Examination of material evidence on site
- Assessing computer records, replay and memory access from GPS, ARPA, electronic charts and machinery and other data records
- Downloading and analysing VDR
- Examining course recorders and other automatic date recording devices
- Obtaining certificates, such as registry, competency, machinery, class and safe manning
- Obtaining weather forecasts and actual weather reports, tidal reports and other ephemeral data
- Conducting specialized studies if necessary
- Looking at organizational factors
- Identifying conflicts in evidence
- Identifying missing information

During the course of gathering evidence, cross-checking the evidence is imperative. Documents investigators collected must contain the same information from different sources, content of witness interviewing must agree with and corroborate physical and electronic evidence. Once all the evidence has been collected, investigators should consider having a peer group review, which enables others to offer an opinion on evidence considered necessary or missing at the early stage of investigation. As the investigation progresses, investigators can remove information irrelevant to the accident.

6.3.2 *Human evidence: Witness interviewing*

Human evidence is the most volatile form of evidence and interviewing is a tool of retrieving human evidence. Interviewing is a difficult skill that demands good technique and practice. PEACE (acronym of Prepare and plan, Engage and explain, Account, clarification and challenge, Closure, and Evaluate) is one of cognitive interviewing techniques, aiming to retrieve as many details and facts in the context of reinstatement. However, in the case of marine accident investigation, it is difficult to apply all of this method because the interview time on site is limited, and in many cases interviews cannot be expected to be in quiet places. This section introduces a useful part of PEACE which can be used for marine accident investigation. In this chapter, witnesses include eyewitnesses, operators, shore managers and others related to the accidents.

Prepare and plan

It is beneficial to identify witnesses and produce a short plan which covers topics and questioning areas prior to on-site investigation. Approach to the interview varies according to the witness. Persons directly involved in the accidents are key witnesses and are asked about all the human factors and events that contributed to the accidents. Crew, company personnel, port officials, equipment designers and emergency response personnel are asked about their relevant topics.

Prior to the interview, available evidence such as AIS records, sailing directions and meteorological data should be gathered, analyzed and assessed. Looking at whether similar accidents have occurred in the past and analyzing them will help investigators prepare for conducting on-site investigation and producing safety recommendations.

Milne and Bull (1999) state the key points of conversation management as follows: (1) The interview is a conversation with a purpose. (2) The interviewer must have some form of control. (3) In order to control, interviewer must carefully prepare and plan for the interview. (4) Preparation includes reading, analyzing and grasping existing evidence, listing key points to prove and preparing a general framework of questions.

Engage and explain

This stage is the opening of the interview. Investigators introduce themselves, an assistant and an interpreter and greet the witness by name.

It can be helpful to start with neutral questions not related to the accident, behaving in a natural manner, not making the interview seem artificial. There are four items to be clarified at the beginning of the interview. (1) investigator's power and obligation, (2) the purpose of the interview, (3) interviewee's rights, and (4) consequences of the information obtained by interviewing.

When using a digital recorder, investigators must ask permission to use it. The digital recorder allows an interviewer to concentrate on witness and interview plan. It has a disadvantage as well in that the interviewer becomes heavily reliant on it and when the recorder happens to put be under the chart and other documents conversations cannot be recorded.

The investigator must record the following items before and after the interviewing. Before the interviewing: Name of ship; date of the accident; date, time and place of the interview; name of the investigator; name of the interviewee together with rank, job title; name of anyone else present at the interview. After the interview: The time when the interview ends.

Account, clarification and challenge

This stage is the main part of the interview, and can be useful to utilize some of cognitive interviewing techniques. These are some of useful cognitive interviewing techniques (Fisher and Geiselman 1992).

(1) Telling the witness to actively generate information and not wait for the interviewer to ask questions.

(2) Asking open-ended questions.

Open-ended question is the best type of question for evidence gathering. It can avoid the risk of imposing the interviewer's view. These are examples of open-ended question: Tell me everything, explain to me, describe it in as much details as you can. On the other hand, a closed question is the second-best type of question because it can only produce a narrow range of information and incorrect responses compared to open-ended questions. However, an interview cannot be completed by using only open-ended questions, so these two types of question should be mixed in order to retrieve more accurate and detailed human evidence.

(3) Listening actively.

(4) Not interrupting.

Cognitive interviewing technique cannot be used on uncooperative witnesses and those who have language barriers. The interviewer can facilitate the evidence gathering using the chart, AIS records related to the accident and by letting the witness draw a picture. In the case of a collision, the interviewer can obtain detailed information more accurately from the witness by moving the place of the interview to the bridge, listening to the situation of the ship maneuvering, the relationship with other vessels and other relevant topics at the time of the accident. Sound records of VDR can help the witness retrieve accurate and detailed memory about the decision-making process of the operator.

In this stage, the human factors related to the accident should be covered, and then these human evidences must be verified by cross-checking with documentary evidence, physical evidence and electronic evidence. Parts of human factors are as follows (IMO 2014a).

People factors
- ability, skills, knowledge
- mental condition, emotional state, operational readiness
- medical fitness, drugs and alcohol
- sleepiness and fatigue
- work and rest hours in the last 7 days
- level of situation awareness
- activities prior to accident
- assigned duties and actual behaviour at the time of accident
- social interaction with people in workplaces

Organization on board
- allocation of tasks and responsibilities
- composition of the crew and manning condition
- workload and complexity of tasks
- procedures, standing orders and night orders
- internal and external communication
- management and supervision onboard
- onboard training and drill related to the accident
- teamwork or BRM
- planning of voyage, cargo loading/unloading, maintenance

Working and living conditions
- ergonomics at local workplaces
- ship motion, vibrations, noises, temperature, humidity

Ship factors
- design including ship and equipment
- state of maintenance
- alteration of design
- characteristics of cargo such as combustible, toxic, oxygen deficient substances

Shoreside management
- policy on recruitment
- safety policy and its commitment by management
- safety culture
- port scheduling
- ship-shore communication

Environment
- sea and weather conditions
- level of traffic density
- ice conditions
- characteristics of waters such as narrow channel, waters with dangers

External influences
- regulations, surveys and inspections including ones of international, national, port, classification societies, industrial and others.

Closure

After completing the interview, the interviewer thanks the witness for the evidence provided and makes the witness feel important to the investigation. The interviewer can encourage the witness to report to the interviewer the information not mentioned during the interview which was recalled after the interview has been terminated.

Evaluate

Complete summaries of the interview are required and whether further investigation is needed is decided. Therefore, it is necessary to compare the content of the interview with other evidence.

Use of an interpreter

Common language aboard the ocean-going vessel is English language. As not all crewmembers understand English, investigators often employ an interpreter for interviewing. There are points to keep in mind when using interpreters (IMO 2014b). (1) Provide a list of common nautical and technical terms to allow the interpreter to prepare themselves. (2) Allow double or more estimated interview time. (3) Explain an interview technique of free recall to the interpreter. (4) Use short sentences in order to allow the interpreter to interpret. (5) Instruct the interviewee to talk in short segments in order to allow the interpreter to interpret. (6) Instruct the interpreter to translate everything that is said. (7) The interviewer should address the witness directly, not talk through the interpreter. (8) After the interview, ask what the interpreter's impression of the witness is.

Interviewing children

Investigators may encounter cases requiring a child's interview as a witness to an accident. Unlike adults' interviews, Fisher and Geiselman (1992) state points to keep in mind as follows:

(1) The answer in the child's interview is greatly affected by the behavior of the interviewer himself and how to ask questions.

(2) For the first few minutes before starting an interview, the interviewer needs to create a good relationship with the child. For example, topics such as friends, family, favorite games, toys, movies and TV shows are turned to.

(3) Because language skills are underdeveloped, when asking a child, avoid long sentences, ask short questions and use words that match the child's age.

(4) Children like closed questions. In the early stage of the interview, after acquiring a wide range of responses in the child's memory with an open-ended question, the interviewer should ask them with a closed question on related topics.

(5) Children may misunderstand the interviewer's intentions and fabricate answers if asked repeatedly. Children should be notified at the beginning of interviews that they should tell only the truth.

(6) Children are not used to explaining things in detail. It is effective to give a child a blank sheet, draw a situation at the time of the accident, or use a method such as recreating the movement of the ship after giving the child a ship model.

6.3.3 Physical evidence

Once physical evidence is removed, it can never be precisely put in the original position. It is necessary to pay attention to the following points (IMO 2014b). (1) Evidence should not be removed until witness interviewing is completed, since the accident site can stimulate a witness's memory. (2) Before removal, physical evidence should be measured and photographed, and its position should be noted on a sketch of the scene. (3) When removing, care is necessary in order to avoid damaging impact marks and fracture surfaces. Damaged or fractured items of evidence should be protected against further damage. For technical examination, delicate parts of items should be padded, boxed, labelled and sent to research institutes. (4) All physical evidence should be catalogued, documented or photographed. (5) Agreement for the removal must be made with other interested parties prior to any action.

If investigators encounter damaged equipment or a missing part of the vessel, it is useful to look at something similar or the same type of vessel in order to grasp the overall picture. Studying system diagrams or design drawing helps them to understand what it should have been.

Bodies

Toxicological testing of drugs and alcohol is necessary for accidents in which central liveware of human factors was a contributor. In cases of occupational casualties, a post mortem of a dead body can reveal important evidence, such as types of toxic gasses and oxygen deficiency.

6.3.4 Documentary evidence

Documentary evidence provides important data and should be preserved and secured as methodically as physical evidence. Documentary evidence includes log books, equipment readouts, course recorder traces, paper or electronic charts, licenses, records such as drill and training records, certificates, photographs, videotapes, safety management manuals and various kinds of procedures. The investigation plan should include the identification of records to be collected and the people responsible for their collection. Ship's records should be cross-checked by such external sources as Vessel Traffic Service tapes, harbor control tapes or logbooks, terminal records and custom records. Documentary evidence, especially one related to ISM Code, provides important clues to contributing factors and organizational factors of the accident.

Environment

Environment or weather condition is a crucial element of decision-making by an operator in collisions, groundings, capsizing, cargo shift, personal injury and evacuation that followed. There are three types of evidence: The forecast weather, the weather reported by witnesses, and the actual weather. Since the weather information obtained from ship's logbook or operator sometimes does not agree with other sources, it is better to obtain information from as many sources as possible. Water temperature is needed when assessing survival factors in cases of missing or man overboard. By acquiring detailed and accurate data on ocean currents, tidal currents, wave height, wind force and wind direction in combination with the ship's track retrieved by AIS or VDR, there is a possibility that the process leading to a capsizing can be reproduced by scientific method.

Photography

Photograph is not only critical in providing evidence or verification but also in providing a permanent visual record of the accident scene. IMO (2014b) recommends a digital camera with more than 35 mm camera. When taking the scene, sequence should follow from general to specific: Firstly overview, secondly mid-range, 3 to 6 m from point of interest, thirdly close up, 1.5 m or less, and then macro range with scale if possible.

Investigators should know that a number of accidents are photographed or filmed by passengers, bystanders and CCTV; those photographs and films may record progress of the accident, the actual weather conditions, the state of the vessel, fire extinguishing and evacuation activities.

In cases of sinking of the vessel caused by lack of stability, previous photographs or historical photographs of the vessel are required in order to analyse the stability before the accident, find any modifications, changes in structure and equipment and movement of cargos. Remote photographs using a remotely operated vehicle (ROV) are employed for the purpose of wreckage photographs.

Photographs are often used to record the computer modeling, simulator reconstruction and tank testing for reproducing the situation of the accident. Those photographs can provide investigators with a record of its observation. A photograph taken in low light levels requires the camera to be set on a "low light", "night" or "delay" setting, in reconstruction of the situation of coast lighting and lamps seen from the location where the master made statement that background lights obstructed the navigation lights of approaching vessels.

6.3.5 *Electronic evidence*

As electronic data is perishable evidence, it should be replayed and downloaded as soon as possible after the accident. If investigators are not properly trained, employ experts to recover electronic evidence. Computers with an Inmarsat earth station terminal can store much of what happens at sea. Investigators should know or seek technical assistance in order to access memory. Most fishing boats are equipped with a GPS whose data can be retrieved by using a Random Access Memory (RAM) card that is specific to the type and manufacturers. Most ocean-going vessels are equipped with VDR, AIS, ECDIS and an Electronic Chart System (ECS). In both fishing boats and ocean-going vessels, collecting data requires special expertise or intensive training.

VDR

There are two kinds of VDR: Simplified Voyage Data Recorder (S-VDR) and VDR. S-VDR is a VDR associated with vessels of over 3,000 gross tons constructed before 1 July 2010, although some vessels that fit the requirement may be exempted by their administration. VDR is required for all passenger vessels and all vessels of 3,000 gross tons or more on international voyages. S-VDR and VDR have such common input as date and time, ship's position, speed, heading, bridge audio obtained from microphones and communications audio from VHF communication. Other data, such as radar, echo sounder, main alarm, rudder order and response, engine order and response, hull opening status, watertight and fire door status and wind speed and direction become options for S-VDR.

The availability of the VDR read-out of the time slot related to the accident is the greatest interest of the investigator. New performance standards of VDR were enacted by IMO (2012), recording period of fixed recording medium was extended from 12 hours to 48 hours, and long-term recording medium was newly established to last 720 hours. If the VDR read-out of the vessel related to an accident is not available, an alternative way is to check other vessels equipped with VDRs which were navigating in the vicinity of the accident site and could provide additional information.

Before obtaining the information recorded on VDR, investigators are recommended to get confirmation from the ship owner and the master. IMO (2002) regulates the ownership and recovery of VDR as follows.

Ownership of VDR information: The ship owner will own the VDR and its information. The owner should make available and maintain all recording instructions necessary to recover the recorded information.

Recovery of VDR information: Recovery of the VDR information should be undertaken as soon as possible after an accident by both the investigator and the ship owner. As the investigator is very unlikely to be in a position to do so, the owner must be responsible for ensuring the timely preservation of information. In case of abandonment of a vessel, the master should take necessary steps to preserve the VDR information until it can be passed to the investigator.

Custody of VDR information: During an investigation, the investigator should have custody of the original VDR information.

Read-out of VDR information: The investigator is responsible for arranging downloading and read-out of the information and should keep the ship owner fully informed. In some cases, the assistance of specialist expertise may be required.

Access to VDR information: A copy of the VDR information must be provided to the ship owner at an early stage in all circumstances.

AIS

AIS is a shipboard broadcast transponder system where vessels continually send the ship's name, position, course, speed and other data to all other nearby vessels and shore authorities on a common VHF radio channel. AIS messages include static data which is programmed at the time of setting up AIS units, dynamic data and voyage related data. The static data involve maritime mobile service identity (MMSI), IMO number, length and beam, type of ship and location of position fixing antenna on ship by aft of bow, port or starboard of center line. The dynamic data are derived from interfaces with the ship's GPS and other sensors, and involve the ship's position, date and time, speed over ground (SOG), course over ground (COG), heading, navigational status and rate of turn. Voyage-related data is entered manually by the crewmember, and includes the ship's draft, hazardous cargos, destination and estimated time of arrival.

ECDIS and ECS

A navigational electronic chart system is a general term for all electronic equipment that is capable of displaying a ship's position on a chart image on a display. There are two classes of navigational electronic chart systems: ECDIS and ECS. ECDIS meets chart carriage requirements on SOLAS Convention, whereas ECS does not meet that requirement. ECS may be able to use either official navigational charts or other charts produced privately and can have functionality similar to ECDIS. ECDIS and ECS data include date and time, position, SOG, COG, planned route and documentary data. Software of ECDIS and ECS are based on a Microsoft Windows operating system. Extracting data requires CD, DVD or USB thumb drive.

GPS

GPS is a satellite navigation system, and more than two dozen GPS satellites send precise timing signals by radio. A GPS receiver can accurately determine its location by longitude, latitude and altitude in any weather, day or night, and anywhere on earth. Differential GPS (DGPS) is a method of improving the accuracy of a ship's receiver by adding a local reference station in order to augment the information available from the satellites. GPS data includes date and time, position, SOG, COG, planned route and track. Volatile memory equipped in older types of GPS cannot be read after immersion. Non-volatile memory can be read after immersion; if immersed, the circuits are rinsed with fresh water and the device is delivered to a technical institution capable of retrieving information from such devices. If GPS devices are operating, it is possible to transfer the data to the device's memory card which is obtained by a manufacturer.

Shore based recordings

Ship traffic management systems, such as VTS and VTIS, can record and store radar image, digital and AIS information. The information is routinely recorded by many coastal states and commercial organizations. In ports, security cameras may record video footage related to the accident. Communication records by VHF or telephone and weather records are often stored in addition to this footage.

Mobile phone records concerning the time and a triangulated position can be helpful when no witnesses or other evidence are available. In order to obtain mobile phone records, investigators need to ask judicial authorities.

6.3.6 Specialist testing and survey

Many accident investigation authorities clarify contributing factors by employing external research institutions, especially to analyze issues related to hardware. The institutions conduct testing of faulty equipment, testing of products against standards, metallurgical testing of fractured components, microscopic and trace element examination, such as paint samples, as well as electronic data recovery, fatigue and impairment analysis, underwater surveys, stability calculation and simulation of sinking, testing of atmosphere and progress of accidents like fire, explosion, death by toxic gasses or oxygen deficiency. Autopsies and drug and alcohol identification are usually undertaken by law enforcement agencies. In order to ensure impartiality and credibility of the testing and investigation, specialists or institutions that are not related to the accident nor the manufacturer of the equipment are selected.

In the case of sinking after collision or sinking due to lack of stability, the investigation of the sinking vessel is important from the viewpoint of finding contributing factors and mitigation of damage to lives, property and environment. Underwater surveys are conducted as follows.

Locating the wreck

The coast guard or navy, in most countries, conducts search and rescue operations for missing people. If salvage of sinking vessel is decided, a salvage company finds the location of the vessel. When approximate position of the wreckage is found by tracing the Emergency Position Indicating Radio Beacon (EPIRB), oil slick, debris or snagging fishing gear, an accurate position is determined by using side scan sonar sweeping a large area of sea bed. Investigators need to maintain close contact with the authorities and salvage companies.

ROV surveys and diver surveys

Detailed survey after determining the location of the wreck is undertaken by ROV or divers. Benefits of ROV surveys and cautionary note for diver survey are shown below (IMO 2014b).

(1) Virtually no limitation on depth of operation.

When the depth of water becomes greater, the bigger ROV can be used, and in shallow waters, a small ROV is used.

(2) Unlimited operation on the sea bed.

The bigger ROV have manipulation arms or other tools to retrieve small artefacts and important evidence. The ROV has a camera and investigators can direct the ROV driver to the location they want. The small ROV can gain access to the wreck, entering enclosed spaces.

(3) No risk to divers.

Divers in scuba gear can stay on the bottom for 20 to 30 minutes at a time for shallow water less than 20 m. For deeper water and complicated works, surface supply divers or hard hat divers are required. When divers are engaged in search and recovery, they should be interviewed in order to determine what they saw and what they tampered with. For instance, when recovering the body of a fisherman out from the wheel house under water, a diver might have changed the position of the engine throttle. Investigators must understand the limitations of the dive time, effect of visibility and underwater current, and brief fully as to what is required on the investigation. Divers should be given as much information on any hazards of the wreck as possible. Depending on the depth of operation, platform, supervisor, relief divers, and decompression arrangement are required.

6.4 Assessing the evidence

Prior to analysis, evidence collected should be assessed. The quality of the accident investigation is highly dependent on the quality of the evidence collected by the investigators.

Electronic evidence, such as VDR, AIS, ECDIS, instrument reading, etc., are, in general, objective, accurate and trustworthy and can greatly contribute towards finding the event chain and contributing factors of an accident. However, the investigators need to understand the limitations of these evidences. The radar image on a VDR readout is an image projected every 15 seconds, during which the movement of the ship is not recorded. The position of the ship indicated by the VDR readout is the same as the position of the GPS antenna, and if the person in charge does not input the position of the antenna accurately as the initial information, the VDR data lacks accuracy and errors are present in the position of the ship itself in AIS tracks.

Human evidence, such as content of interviews from witnesses, depends on the number of days elapsed from the day of the accident occurrence to the execution of the interview, and on the skill of the investigator performing the interview. Therefore, the value of human evidence is lower than that of electronic evidence.

Strauch (2004) lists the following five factors which influence the value of written documents that investigators collected: (1) quantity, (2) frequency and regularity, (3) temporal closeness, (4) reliability, and (5) validity.

Quantity

For certain information, the more evidence investigators have, the higher the value of the evidence. In other words, if the content of information obtained from different kinds of evidence converges to one, that information is worth looking at and contributes to the analysis.

For example, if electronic data, such as VDR readout, AIS records, etc., were not available and only one oral statement from interviewing an operator was obtained, the contents of the statement would be of low value as evidence. If an accident occurred due to not securing the starboard side door of the cabin while a high-speed passenger ship was sailing, interviews were conducted by a large number of passengers aboard the ship on the opening and closing status of the starboard side door, and all of them interviewed stated that "the door of the starboard side was opened, not shut while underway", the said information would be valuable as evidence.

Frequency and regularity

If the condition of things is measured frequently and periodically, investigators can know the transition of the accurate change of the condition. Measurements need to be as frequent and regular as possible. In this point, the ship's tracks retrieved by AIS can hold high value because all of the ship's data, including time, position, speed and true course, are sent and received by the ship and land stations, respectively, at regular intervals. In an accident occurring in an enclosed space during an unloading operation, if operators regularly measure the oxygen concentration in the cargo hold before entering there, the operators can know the exact change in oxygen concentration during that time periods, helping not only them but investigators to assess risks prior to commencing the work.

Temporal closeness

The value of the data becomes closer to the actual value at the time of the accident when the data is obtained both immediately before the accident or immediately after the accident. For instance, regarding training records of life boats onboard the ship, if the registration of the life boat training record is just before the accident, it is valuable in evaluating the effectiveness of that training.

Reliability

Reliability indicates that consistency exists in measuring means: Those with reliability change only slightly, but those that are not reliable change substantially. Although the weight measured by the calibrated body weight scale is small with the exception of the subject's daily changes, when measured with a scale that is not calibrated every time, the change becomes large and it can be said that the reliability is lacking. In an interview, when the person has not actually witnessed the accident but dictates indirectly what he heard from others or that the person has not witnessed the fact but spoken on guess, the content of the oral statement will be of low reliability. In an interview with witnesses, the content may be inconsistent, but it cannot be concluded that the witness's dictation content is all wrong due to lack of consistency.

Validity

It shows the relationship between the measurement value of the parameter and the actual value, and if the two values are close, the validity is high. When checking certificates of competency, attention is needed in order to make sure that they are authentic and issued by the appropriate administration. It can be said that the validity is high in the following cases. When the master's certificate of competence was the original and the master's skill met the qualification requirements of the ship's master set by the management company, the master's statement would be accompanied by validity. If a person who expressed opinions on a certain fact was an authority in the field related to the fact, then his statement would be regarded as valid.

6.5 Conclusions

Performing scientific accident investigation or evidence-based investigation is a prerequisite for systematic accident prevention. The principle of collecting evidence is to collect perishable evident first, and then the rest. Perishable evidence includes human evidence, electronic evidence such as VDRs, AIS, GPS, physical evidence such as pre-rescue scene, metallic fracture surfaces and atmosphere in cargo tanks and enclosed spaces. There have been accident investigation reports issued which conclude the process of accident occurrence, contributing factors and safety recommendations without enough evidence and assessment of evidence. Safety recommendations obtained in these reports are out of place and ineffective in preventing recurrence of accidents. As pointed out by ICAO, the quality of the accident investigation report greatly depends on the quality of the investigator, therefore, it is necessary for the investigators to master the methods of marine accident investigation.

Chapter 7

Analysis Methods

7.1 Introduction

Although there are many different types of charting techniques which can graphically depict the accident event chain and contributing factors, IMO adopts Event and Contributory Factor Charts (ECFC) as an analysis method (IMO 2014b). In this chapter, analysis process, ECFC and its relation to a marine accident investigation reports are introduced mainly according to the IMO relevant resolutions and publications.

7.2 Analysis process

7.2.1 Terms and concept

IMO (2014a) defines the terms used in the analysis. The words "event" and "condition" frequently appear in the terms. These terms are related to each other: Certain conditions cause certain events, and then the events lead to new conditions, and the transition reiterates.

Event: An action, omission or other happening. Events are generally based on solid facts.

Accident event: An event that is assessed to be inappropriate and significant in the sequence of events that lead to the incident and accident.

Casualty event: The accident and incident, or one of a number of connected accidents and/or incidents forming the overall occurrence.

Contributing factor: A condition that may have contributed to an accident event or worsened its consequence.

Safety issue: An issue that includes one or more contributing factors and/or other unsafe conditions.

Safety deficiency: A safety issue with risks, for which existing defenses aimed at preventing an accident event, and/or those aimed at eliminating or reducing its consequences, are assessed to be either inadequate or missing.

To clarify the concept of the terms, a comparison of the terms defined by IMO, with concept of hazard and accident defined by ISO/IEC Guide 51, is shown in Fig. 7.1. A harmful event is equivalent to an accident event, and harm is the same as a casualty event. An unsafe condition means a hazardous situation. A contributing factor is related to the progress from the hazardous situation to the harmful event, and from the harmful event to the harm. Safety deficiency means the flaws in defenses which are embedded in the system to prevent the progress of the harmful event and the harm. Safety deficiency is on the main topic of producing recommended safety actions.

7.2.2 Steps of analysis

IMO adopts the concept of the SCM, but it points out the SCM's lesser emphasis on technical failure which does not result from an operator's

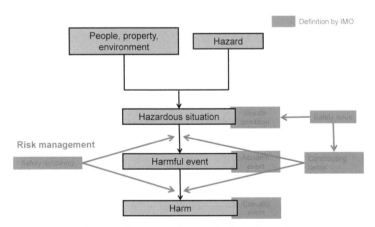

Fig. 7.1: Concept of terms defined by IMO.

error and then it incorporates technical failure in analysis process. These are the steps to follow during the process of collecting evidence and analyzing collected evidence (IMO 2014b).

(1) Develop an event-chain.

During on-site investigation, investigators collect evidence from interviewing witnesses, VDR and other electrical equipment, damages on vessels and shore facilities and documents that relate to the accident, then complete the timeline of the events, arranging them chronologically from left to right.

(2) Identify casualty and accident events.

IMO model describes collision, grounding, flooding, capsizing, fatality, loss of control, and structural failure as casualty events as a sample. As a criterion for judging whether it is an accident event or not, the investigator decides whether another well-run ship will cause a similar event under similar circumstances. If the answer is negative, then the event is an accident event.

(3) Analyze each accident event to find contributing factors to individual actions, human error and technical failure mechanisms.

This stage uncovers the mechanisms and error types which best describe human error and technical failure. CREAM, of which classification includes action, observation, interpretation and planning or intention, and GEMS, which classifies unsafe acts and error types, can be used. In addition, regarding cases involving collisions, Endsley's situation awareness model can help in identifying human errors related to situation awareness.

(4) For each individual action and technical failure, analyze the local workplace factors which may have had an influence, including any risk controls that failed or were not in place.

This stage looks into conditions behind each accident event existing onboard the ship by using why-why analysis in conjunction with the SHEL model. Some contributing factors may be presumptive at this stage and more evidence to support it is needed.

(5) Analyze further to find organizational factors that underlay the events.

As with step (4), methods to find contributing factors at an organization are the same: Using why-why analysis and the SHEL model. Those contributing factors are safety issues and safety deficiencies which provide a basis of recommended safety actions.

The stop rule is established; investigation is finalized when no organization can address the contributing factors discovered by the analysis.

IMO recommends using ECFC to graphically illustrate the steps and the results of the analysis shown above.

7.3 Event and Contributory Factors Charts

ECFC (DOE 1999) is useful in identifying contributing factors and graphically depicting the triggering conditions and events necessary and sufficient for an accident to occur. It is a graphical display of the accident chronology and is used primarily for compiling and organizing evidence to illustrate the sequence of the accident events. When other analytical techniques such as change analysis are completed, they can be incorporated into this chart.

There are several benefits when using this analysis method.

(1) Illustrate the sequence of events leading to the accident and the contributory factors.

(2) Let investigators get to know necessary data collection and analysis by identifying information gaps and become good communication tool among investigators.

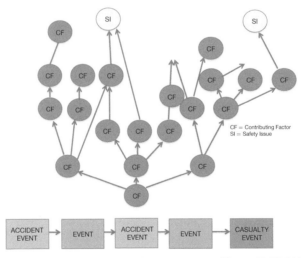

Fig. 7.2: Structure of Event and Contributory Factors Charts (IMO 2014b). SI and CF are acronyms for Safety Issue and Contributing Factor, respectively.

(3) Providing a structured method for collecting and integrating collected evidence.

(4) Present information regarding the accident that can be used to guide report writing.

There are some weak points as follows.

(1) Not available for systemic accident, even if applicable, graphical depiction of events and contributing factors become complicated and difficult to understand.

(2) Cannot depict an accident which involves multiple organizations.

(3) Rely on such unambiguous definition of accident event as being inappropriate and significant in the sequence of events that led to the accident.

7.4 Guidelines to produce ECFC

IMO (2014b) explains how to produce the ECFC. The process starts chronologically from left to right, and in more complex accident cases, secondary and miscellaneous events can be added to the ECFC by inserting a line above or below the primary sequencing line. When presumed conditions are included, appropriate weight should be given in the final analysis.

How to illustrate events

(1) Events are described by a short sentence with one subject and one active verb.

(2) Each event rectangle block has the time and date of the event.

(3) Event rectangle blocks are connected with solid arrows.

(4) Presumptive events are shown by dashed rectangles and connected with dashed arrows.

(5) If more than one event lines are used, those lines are depicted in parallel horizontal or vertical lines.

How to illustrate contributing factors

(1) Contributing factors are enclosed in oval.

(2) All contributing factors are connected to the preceding and subsequent events by arrows.

(3) Presumptive contributing factors are shown by dashed ovals and connected with dashed arrows.

7.5 Construction of a marine accident investigation report

The steps of analysis and graphical chart shown in the previous section can facilitate the creation of marine accident investigation reports. CIC Code 2.12 prescribes the structure of a marine accident investigation report as follows.

(1) A summary outlining the basic facts of the marine accident or incident stating whether any deaths, injuries or pollution occurred as a result.

(2) The identity of the flag state, owners, operators, the ship management company, and the classification society.

(3) The details of the dimensions and engines, if necessary.

(4) A description of the crew, work routine and other matters, such as experience and time served on the ship.

(5) A narrative detailing the events and surroundings of the marine accident or incident.

(6) Analysis and comment on the causal factors, including any mechanical, human error and organizational factors.

(7) A discussion of the safety investigation's findings, including the identification of safety issues, and the safety investigation's conclusions.

(8) Where appropriate, safety recommendations aimed at preventing future accidents and incidents.

As written above, a marine accident investigation report must contain factual information, analysis, conclusions and recommended safety actions. There are close relationships among them: Factual information leads to analysis, which leads to conclusions that eventually lead to recommended safety actions.

Factual information: Factual information contains all the information which is essential for the development of analysis, conclusions and recommended safety actions, and is written in chronological order. Supporting documents, such as photographs, AIS records, VDR readouts, data and conclusions derived from scientific experiment and studies are included or put in the appendix. Corrective actions taken by the organization that are related to the accident, as well as a relevant extract of regulations, are also included.

Analysis: Analysis provides a logical link between the factual information and the conclusion. Basically, the marine accident investigation report is similar to a scientific article. However, the former often includes gaps in factual information, mainly due to the limitations of the investigation with regard to available time to finalize the report and budgetary constraints. The gap should be filled in by extrapolation from available information using logic based on knowledge of the shipping industry. Logical reasoning should be used to make hypotheses which are discussed and tested against evidence. Speculation and what is not known should be clarified.

Conclusions: They maintain the same degree of certainty written in analysis. For instance, if "probable" is used in analysis, then the same word is used in conclusions.

Recommended safety actions: ICAO (2001) stipulates the definition of a safety recommendation in Annex 13 to the Convention on International Civil Aviation as follows. "A proposal of the accident investigation authority of the State conducting the investigation, based on information derived from an investigation, made with the intention of preventing accidents or incidents." Safety recommendation in no case has the purpose of creating a presumption of blame or liability for an accident or incident. In addition to safety recommendations arising from accident and incident investigations, safety recommendations may result from diverse sources, including safety studies. Safety studies involve examining a particular safety issue, like the effectiveness of the snap-back zone painted on a fore deck, accident trend from accumulated accident investigation reports, or other accident data.

Recommended safety actions describe the safety issues and safety deficiencies and provide justification for the safety actions. Each recommended safety action includes a specific addressee in order to ensure the implementation of the actions. If safety actions were taken by the recipient, safety actions taken are excluded from the recommended safety actions, and written under a separate heading. During an accident investigation, safety deficiencies unrelated to the accident are often found, and safety recommendations need not be confined to the cause of the accident, but should be related to matters covered in the investigation.

Safety recommendations should not increase the risks and put great hazards elsewhere in the ship's operation. The efforts and cost of safety measures should be balanced with the severity of the risks posed.

When safety deficiencies are thought to present serious risks to lives, properties or environment, early alerts or interim safety recommendations should be issued. If safety recommendations are not made, the recommendation section can be excluded; instead, appropriate heading of safety actions taken should be made (IMO 2014b).

Follow up: Since recommended safety actions are not mandatory for the recipients, some marine accident investigation authorities publish an annual report listing the outcome of their recommendations made for the purpose of preventing similar accidents. When an accident report is produced by an organization or ship management company by itself, monitoring and review of recommended safety actions is critical to prevent recurrence.

7.6 Conclusions

For accident investigators, report writing of marine accident investigation is the final product and materialization of on-site investigation, analysis and conclusions. Contents from Chapter 2 to Chapter 7 will satisfy their demands. However, these do not answer frequent questions that author used to be asked as to how to prevent accidents. Chapter 8 to Chapter 11 provide a specific approach of how to prevent and reduce the number of marine accidents and incidents.

Chapter 8

Visualization of Weakness in the System

8.1 Introduction

The SCM has unsolved issues, and this chapter discusses those issues while illustrating the location of holes and latent conditions with a new accident model that was developed by the Late 1 SCM. To determine the location of holes in local workplaces and organizations, risk management process and safety management systems are used (Fukuoka 2016a, 2016b).

It is reminded that making clear an event-chain of an accident is an important and indispensable process in using any accident model, and it is a prerequisite to understanding characteristics of three different categories of accident models which can be applied to own field.

8.2 Background

To understand how accidents occur and to develop preventive measures against accidents, various kinds of accident models have been developed. The accident model is divided into three categories: Sequential accident models, systemic accident models, and epidemiological accident models.

The sequential accident model is represented by domino theory, and is applicable to accidents that have clear cause-effect links. However, this model is not suitable for explaining accidents in complex systems in modern society.

The systemic accident model is represented by the functional resonance accident model (FRAM), which rejects cause-effect links and regards the accident as an emergent phenomenon (Hollnagel 2004). However, the FRAM is not widely used in practice (Reason et al. 2006).

The SCM (Reason 1997) is categorized as an epidemiological accident model, which explains accidents in complex systems and is used in the marine, aviation, railway, road transportation, and medical fields (Anca 2007). Using the marine and aviation fields for the theoretical background, the SHEL and Reason hybrid model, which combines the SHEL model with the SCM, was developed and used by national safety investigation authorities (IMO 2000a). In addition, the Tripod-Delta model, a sociotechnical system model that is based on the SHEL model (Grech et al. 2008), and the human factors analysis and classification system (HFACS) (Wiegmann and Shappell 2003), which is based on the SCM, have been developed as accident models. During on-site accident investigations, marine and aviation safety investigation authorities use the SHEL model to collect evidence (IMO 2000c, ICAO 1993).

According to the SCM, a number of defensive layers and associated holes exist between hazards and potential losses. These holes are in continuous motion, moving from one place to another, and opening and shutting. Holes are caused by latent conditions and active failures. Latent conditions can serve both to promote unsafe acts and to weaken defensive mechanisms. No one can foresee all possible accident scenarios. Therefore, some holes in defensive layers will either be present at the time of system establishment or will develop in an unnoticed or uncorrected manner during system operation. When such holes line up in a number of defensive layers, hazards come into direct contact with potential losses and an accident occurs.

In the context of the SCM, potential accidents and losses can be avoided by preventing holes from lining up. This means that when holes that have been lined up as a result of latent conditions are shut, accidents and losses do not occur. Therefore, to take effective and efficient preventive measures, it is important to determine the relationship between latent conditions and the characteristics of holes caused by those conditions. If holes can be visualized and the relationship between holes and latent conditions can be made clear, it is possible to control the occurrence of holes and thereby reduce the number of accidents.

However, the SCM, the SHEL and Reason hybrid model and the HFACS do not consider the location of holes and the relationship between holes and latent conditions.

157

8.3 Holes

8.3.1 Definition of a hole

According to ISO/IEC Guide 51, safety does not mean absolute freedom from risk, but rather freedom from unacceptable risk. Unacceptable risk is that which exceeds the limit of the tolerable region defined by the as low as reasonably practicable principle (ALARP) (IEC 2009). Safety is achieved by reducing risk to tolerable levels. The opening of a hole in a defensive layer is defined as an unacceptable risk in an organization or at a local workplace (Fig. 8.1).

When defining a hole, hazardous situations that could result in an accident are considered. A hazardous situation is one in which there is exposure of people, property, or the environment to one or more hazards as defined by ISO/IEC Guide 51.

The following are definitions of unacceptable risk situations at a local workplace: In collisions, a situation in which two vessels approach within the maximum advance of a give-way vessel; in grounding, a situation in which a vessel enters a no-go area; in occupational casualties while entering an enclosed space, a situation in which a crew member enters an enclosed space without checking the atmosphere.

An SMS in the marine domain is established in accordance with the requirements of the ISM Code. Section 1.4 of the ISM Code outlines the functional requirements for an SMS. An unacceptable risk situation in an

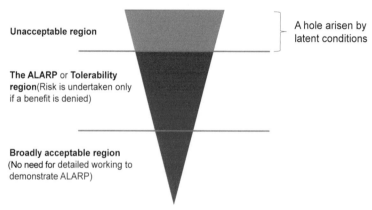

Fig. 8.1: Relationship between the ALARP principle and a hole (Adapted from IEC/ISO 31010).

organization is defined as one in which the functional requirements related to accident prevention as prescribed by the ISM Code are not satisfied.

8.3.2 *How to identify the movement of a hole*

To locate opening and moving holes in defensive layers, observation is made on SMS and risk management processes that are used in an organization or at a local workplace at the time of an accident.

At a local workplace, prior to system operation, risks associated with the system must be reduced until they fall within the tolerable region by taking protective measures and considering the priority order described in ISO/IEC Guide 51. When risks are not reduced to within the tolerable region, this situation is considered equivalent to holes opening in a defensive layer during one of the processes and then moving through the risk management process until an accident occurs.

In an organization, when the functional requirements related to accident prevention as prescribed by the ISM Code are not satisfied, this situation is considered equivalent to holes opening in the SMS defensive layer during one of the processes of the PDCA cycle and then moving through the cycle until an accident occurs.

With regard to the risk management process, terms defined by International Standards 31000 (ISO 2009) and 31010 (IEC 2009) are used. Risk management is considered to be embedded in the procedures for not only collision avoidance actions but also passage planning and enclosed space entry. In the following section, the relation between risk management and procedures is explained. In addition to these procedures, risk management is directly applied to procedures used at the time of accidents, in accordance with the definitions given in International Standards 31000 and 31010.

Figure 8.2 illustrates the processes of risk management (ISO 2009, IEC 2009) embedded in passage planning and the procedures. It is idealistic to consider that the system operates after the level of risks has been reduced to the tolerable region; however, as Reason (1997) states, all possible accident scenarios cannot be foreseen. Therefore, a hole can arise through latent conditions when the system operates.

Figure 8.3 illustrates the PDCA cycle described by the ISO (2008a). The Safety Management System in marine domain is established by the requirements of the ISM Code which is based on the quality management systems of the International Standard 9001 (ISO 2008b).

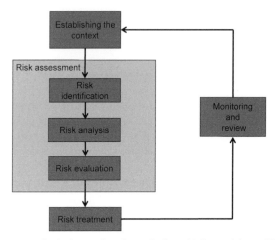

Fig. 8.2: Movement of a hole at a local workplace (Adapted from ISO 31000:2009, IEC/ISO 31010: 2009).

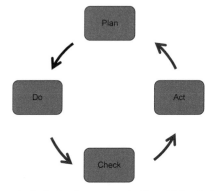

Fig. 8.3: Movement of a hole at an organization (Adapted from ISO 9000).

8.4 Latent conditions

8.4.1 Definitions of latent conditions

Reason (1997) states that latent conditions include poor design, insufficient supervision, unworkable procedures, and lack of training. To define latent conditions, in addition to the results of causal factors in relation to the marine accidents investigated by the JTSB, the concepts of the SHEL model (Hawkins 1987), the IMO/ILO process for investigating human factors (IMO 2000c), the HFACS (Wiegmann and Shappell 2003), and the sociotechnical system model (Grech et al. 2008), as well as the concepts provided by Reason (1997), Swift (2000), Adams (2006), and Parrot (2011), were modified.

In accordance with the IMO/ILO process for investigating human factors, software includes organizational policies, procedures, manuals, checklist layouts and charts. Hardware includes the design of workstations, displays and controls. The environment includes the internal and external climate, temperature, visibility, regulatory climate and other factors that constitute conditions in which people are working. Central liveware includes the capabilities and limitations of the operator. Peripheral liveware includes management, supervision, crew interactions and communications.

8.4.2 10 latent conditions

Considering all related studies and findings mentioned above, the following 10 latent conditions are defined as part of the systematic accident prevention model (Table 8.1) (Fukuoka and Furusho 2016a).

(1) Passage planning
(2) Procedures
(3) Rules
(4) Human-machine interface
(5) Condition of equipment
(6) Environment
(7) Condition of operators
(8) Communication
(9) Teamwork at a local workplace
(10) Management by an organization

Of these conditions, (1), (2), and (3) are liveware-software interactions; (4) and (5) are liveware-hardware interactions; (6) is a liveware-environment interaction; (7) is the central liveware; and (8), (9), and (10) are central liveware-peripheral liveware interactions.

With regard to software, Adams (2006) states that standardized procedures include governmental regulations, checklists, station bills, voyage plans, standing orders of captains and company rules. When investigating various types of marine accidents, it was found that operators at local workplaces were using different kinds of procedures

Table 8.1: 10 latent conditions.

Element of the SHEL	10 latent conditions	Definitions
	Passage planning	Stages of appraisal, planning, execution and monitoring.
	Procedures	Manuals, checklists, station bills, standing orders of captains and company rules.
Software (S)	Rules	The Convention on the International Regulations for Preventing Collisions at Sea, 1972 (COLREGs), International Convention on Standards of Training, Certification and Watchkeeping for Seafarers, International Convention for the Safety of Life at Sea, and local navigation rules.
Hardware (H)	Human-machine interface	Design of work stations, displays, controls and other factors that constitute human-machine interface.
	Condition of equipment	Maintenance of ship and equipment. Condition of stability.
Environment (E)	Environment	Sea and weather conditions, traffic density, geographical features of waters, such as narrow channels, berth facilities and other factors that constitute conditions in which people are working.
Central liveware (L1)	Condition of operators	Physical limitations, physiological conditions, psychological limitations and individual workload management, as well as knowledge, skill, experience, education and training.
	Communication	Communication among the bridge team, between a pilot and the bridge team, or between the bridge and vessel traffic services.
Peripheral liveware (L2)	Teamwork	Roles and responsibilities of the crew, the pilot and other people involved in an accident.
	Management	Functional requirements related to accident prevention as prescribed in the ISM Code, resource management and safety culture.

at the time of accidents. In cases involving a collision, the procedures for collision avoidance were being followed; in cases involving grounding, the procedures for passage planning were being followed; and in cases involving occupational casualties during cargo operations, the procedures for loading or unloading operations were being followed. In addition, Adams (2006) and Parrott (2011) explain causal factors in relation to grounding by using passage planning. Therefore, the standardized procedures were divided into three categories: Passage planning, procedures and rules. Because of this classification, the regulatory climate is categorized into liveware-software interactions. Procedures are defined as the standardized procedures defined by Adams, excluding passage planning and rules (governmental regulations). The definitions of passage planning were based on the concepts provided by Swift (2000) and the IMO (2000b), which consist of the appraisal, planning, execution and monitoring stages. Rules includes COLREGs, STCW Convention, SOLAS Convention and local navigation rules.

With regard to hardware, a distinction is made between human-machine interface and condition of equipment. Human-machine interface includes poor design of work stations, displays and controls. Condition of equipment includes lack of maintenance of ship, engines, equipment and devices. Lack of stability is included in the condition of equipment.

The Environment includes traffic density and geographical features of waters, such as narrow channels. When berth facilities are unfit for vessels and present the possibility of an accident, this is included in adverse environment.

With regard to central liveware, the sociotechnical system model states that an individual includes their physical limitations, human physiology, psychological limitations, individual workload management and experience, skill and knowledge. Education and training are included because skill and knowledge are closely related to education and training (Hawkins 1987).

With regard to peripheral liveware, communication includes that among the bridge team, between a pilot and the bridge team, or between the bridge and vessel traffic services (Adams 2006, Parrott 2011). Teamwork refers to the roles and responsibilities of the crew, the pilot and other people involved in an accident (Adams 2006, Parrott 2011). Management is defined as a situation in which the functional requirements related to accident prevention as prescribed in the ISM Code are not satisfied. In addition, management includes the safety culture defined by Reason (1997).

8.5 How to find a hole at a real accident

8.5.1 Collision

Procedures for collision avoidance are based on COLREGs. They are shown in Chapter 4: Liveware-software, collision and COLREGs.

Collision between the *MEDEA* and the *KOSEIMARU*

The collision between car carrier *MEDEA* and the fishing vessel *KOSEIMARU* is a sample of how to find a hole in an actual case. Figure 8.4 illustrates the tracks of both vessels retrieved by VDR and AIS equipped on the *MEDEA* (JTSB 2012a).

The pilot association

The accident occurred off the East Passage of Nagoya Port, at about 20:30 on 18 March 2010. One factor of the accident was a lack of communication between the pilot onboard the *MEDEA* and the bridge team. A similar accident occurred in Ise Bay on 13 March 2009, when two different pilots of the pilot association had an accident of cargo vessels because they had not shared information on approaching vessels with the bridge teams. Had the pilot and the bridge team onboard the *MEDEA* shared information, it is likely that they could have avoided the situation in which the *KOSEIMARU* passed within the maximum advance of the *MEDEA*. The Safety Management System operated by the pilot association had not afforded enough attention during the processes of check and act concerning sharing information between the pilot and the bridge team since the previous accident. Such a situation indicated that a hole in a defensive layer caused by latent conditions opened during the process of check on 13 March 2009 and moved through the PDCA cycle until 18 March 2010 (Fig. 8.5).

These are preventive measures to be considered by the method of risk reduction:

(1) The pilot association should educate and train pilots about the fact that sharing information about approaching vessels between pilots and bridge teams can prevent collisions.

(2) The pilot association should check the pilots in order to ensure that sharing information about approaching vessels between the pilot and the bridge team is being performed.

The pilot

The pilot detected the *KOSEIMARU* by radar between 20:20 and 20:22, and he recognized that the *KOSEIMARU* was the hazard. He identified the risk of risk management process embedded in the procedures of collision avoidance action. However, he conducted insufficient risk analysis when he thought the *KOSEIMARU* would maneuver away from his intended course, based on her heading and speed that he had acquired from the

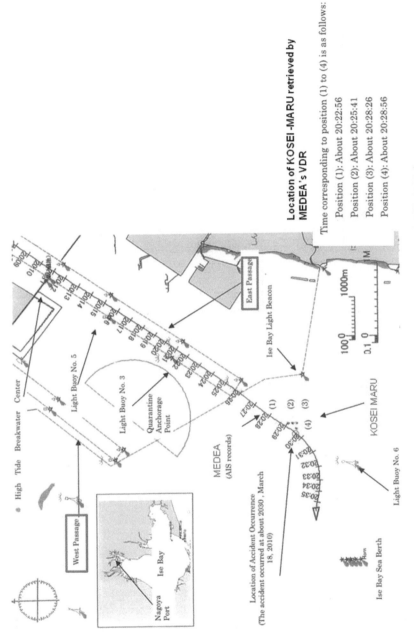

Fig. 8.4: Tracks of *MEDEA* and *KOSEI-MARU* (JTSB 2012a).

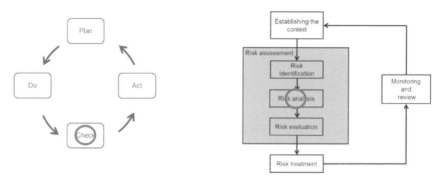

Fig. 8.5: Location of holes of the pilot association (left) and the pilot.

Automatic Radar Plotting Aids (ARPA). He could not foresee during the process of risk analysis that the *KOSEIMARU* would approach within the maximum advance of the *MEDEA*. He did not inform the bridge team of the *KOSEIMARU*'s presence. The pilot did not recognize the necessity of sharing information of approaching vessels with the bridge team. Such a situation in a local workplace indicated that a hole in a defensive layer caused by latent conditions opened during the process of risk analysis (Fig. 8.5). This indicated that the hole in the defensive layer opened at the organization affected another defensive layer at the local workplace between 20:20 and 20:22, and then the hole in another defensive layer moved through the process of the procedures of collision avoidance action until the accident occurred at about 20:30.

If the pilot had shared information on the presence and movement of the *KOSEIMARU* with the bridge team, he could have correctly estimated the risk before the *KOSEIMARU* passed within the maximum advance of the *MEDEA* because the bridge team could have recognized her as a drifting vessel and monitored her movement.

The pilot altered course from 218 to 216 and then 212° when the distance between the *MEDEA* and the *KOSEIMARU* had reduced to 630 m or 0.34 miles. The unsafe act performed by the pilot was that he altered her course from 218 to 216 and then 212° and approached the *KOSEIMARU* under the condition that the *KOSEIMARU* was approaching within the maximum advance of the *MEDEA*.

These are latent conditions on the pilot.

(1) The pilot's initial voyage plan was to navigate the west water of No. 6 light buoy; however, when he noticed that a cargo vessel sailing the west route of Nagoya port was approaching the *MEDEA*,

he decided to proceed via the east water of the light buoy, resulting in a close quarter situation with the *KOSEIMARU*. Alteration of the voyage plan was not conducted by appropriate risk assessment.

(2) After knowing the course and speed displayed by ARPA, the pilot assumed that the *KOSEIMARU* would move away from the planned route of the *MEDEA* and did not systematically observe the movement of the *KOSEIMARU*, the action which deviated from CORLEGs.

(3) The pilot association did not have any rules or guidelines on sharing information about approaching vessels among a pilot and the bridge team.

(4) The whistle push-button at the bridge of the *MEDEA* was arranged on the control panel of the radar, and it was difficult for a pilot to recognize the location of the button at night.

(5) The accident occurred in congested waters in Ise bay, which is one of the most congested waters in Japan.

(6) There was a lack of communication between the pilot and the master of the *MEDEA*.

(7) The pilot did not provide information about approaching vessels, including another cargo vessel and the *KOSEIMARU*, as a result, responsibilities of the pilot and the master were not being performed.

(8) Although the Pilots' Association to which the pilot belonged was alerting attention to fishing boats in Ise bay, the association had not given guidance about pilots' need to share information about approaching vessels with the bridge team. Safety management of the Pilots' Association was inappropriate.

These are preventive measures to be considered by the method of risk reduction:

(1) The pilot should share information about approaching vessels with the bridge team.

(2) The pilot should conduct risk analysis based on sufficient information on the movement of a fishing vessel.

The master and ship management company of the *MEDEA*

The ship management company prescribed on the procedures that the bridge team had to monitor and ensure that the pilot operated safe navigation, but did not prescribe sharing information on approaching vessels with the pilot onboard. The master was not trained in BRM.

This indicated that the SMS of the ship management company could not foresee that there was a possibility of another vessel approaching within the maximum advance of the *MEDEA* if the bridge team did not share information about approaching vessels with the pilot. A hole in a defensive layer caused by latent conditions opened during the process of plan embedded in the SMS on the date created and moved through the PDCA cycle until the accident occurred (Fig. 8.6).

At the local workplace, the master was not aware of the presence and movement of the *KOSEIMARU* until he was informed by the third officer who saw its green light just before the collision. This indicated that he did not identify the hazard. Therefore, a hole in another defensive layer opened during the process of risk identification and moved through risk management process embedded in the procedures of collision avoidance action until the accident occurred.

With regard to the *MEDEA*, four defensive layers and their associated holes were related to the accident (Fig. 8.7).

The master did not sound the whistle when the pilot ordered him to do so in order to attract the attention of the *KOSEIMARU* when the distance between the ships fell to between 0.23 and 0.28 miles. The unsafe act performed by the master was that he did not sound the whistle because he could not find the location of the whistle push-button.

These are latent conditions on the master.

(1) The Safety Management Manual did not stipulate that the bridge team would share information on approaching vessels with a pilot.

(2) The master was not aware of the presence of the *KOSEIMARU* until just before the collision because he was not performing appropriate lookout, which deviated from CORLEGs.

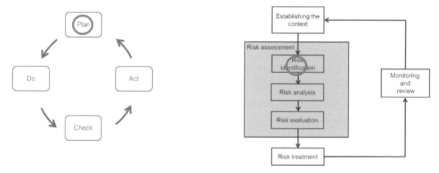

Fig. 8.6: Location of holes of the company (left) and the master of the *MEDEA*.

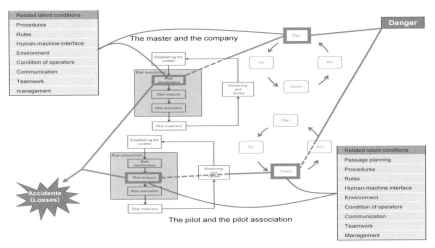

Fig. 8.7: Accident model on the *MEDEA*.

(3) The whistle push-button was installed on the control panel of the radar and it was difficult to ascertain the position of the push-button at night. When the pilot instructed to sound the whistle immediately before the collision, the master could not press the push-button because he could not find it.

(4) The accident occurred in congested waters in Ise bay, which is one of the most congested waters in Japan.

(5) The master did not take BRM training.

(6) The master entrusted the pilot with maneuvering the vessel and did not fulfill the responsibility of the master.

These are preventive measures to be considered by the method of risk reduction.

(1) The management company should establish a procedure to share information about approaching vessels with the pilots onboard.

(2) The management company should train the master for BRM, and provide education and training for the prevention of collision by sharing information on approaching vessels between the bridge team and the pilot.

(3) The master should familiarize himself or herself with the location of the whistle push-button.

Arrangement of defensive layers on the *MEDEA*

When two organizations are involved in an accident, if defensive layers are arranged in series based on time sequence, the SMS defensive layer of the pilot association will be located furthest away from the accident. Next, it is considered that SMS defensive layer of the *MEDEA*'s ship management company, and then risk management defensive layer of the pilot and risk management defensive layer of the master of the *MEDEA* are arranged in the immediate vicinity of the accident.

In the case of this arrangement, although it was confirmed that the hole opened in the SMS defensive layer of the pilot association influenced the procedure of collision avoidance action of the pilot, there is the disadvantage that it is impossible to express the influence of the pilot association on the pilot.

Similarly, the influence of the ship management company on the master of the *MEDEA* cannot be expressed. The master of the *MEDEA* did not share information about the existence and movement of the approaching vessels with the pilots because the ship management company did not prescribe the importance of sharing information with the pilot in SMS and that the master was not provided with the training on BRM.

This case revealed that if the operators belonging to multiple organizations exist simultaneously in a local workplace, it is not possible to arrange each defensive layer in series as drawn on the SCM.

The captain of the *KOSEIMARU*

The captain of the *KOSEIMARU* finished the preparations for fishing operations at the fore deck with working lights while drifting, entered the wheel house, and looked around the surroundings with his naked eyes. The *KOSEIMARU* proceeded at about 6 knots to a fishing ground at about 20:28. He was not aware that the *MEDEA* was approaching on the starboard side because he did not maintain a proper look-out by radar. He did not identify hazards during the process of the procedures of collision avoidance action. This indicated that a hole in a defensive layer caused by latent conditions opened at about 20:28 and the hole moved through risk management process until the accident occurred. The other crewmember was also not aware of the *MEDEA* because he was looking out on the port side.

An unsafe act performed by the captain was that he put helm to the water where the *MEDEA* was taking her course under the condition that the *KOSEIMARU* was approaching within the maximum advance of the *MEDEA*.

These are preventive measures to be considered by the method of risk reduction:

(1) The captain of the fishing vessel should use all means, such as the radar function of an alarm, to detect an approaching vessel in order to maintain a proper look-out.

(2) The captain of the fishing vessel should strictly observe the look-out by radar taking into account that he may not recognize an approaching vessel with naked eyes under the effects of working lights and dark adaptation.

8.5.2 *Enclosed space entry*

Procedures for enclosed space entry are based on the revised recommendations for entering enclosed spaces aboard ships issued by the IMO (2011). It is considered that Sections 4.1 and 4.2 of Resolution A.1050(27) covered risk identification and risk analysis; Section 4.3 covered risk evaluation; Sections 4.4 and 4.5 covered risk treatment; and Section 8.1 covered monitoring and review.

A summary of the revised recommendations for entering enclosed spaces onboard ships is as follows:

(1) Sections 4.1 and 4.2 state that the company should ensure that a preliminary assessment is conducted in order to identify all enclosed spaces on board the ship, and that the preliminary assessment should determine the potential for the presence of oxygen-deficient, oxygen-enriched, flammable or toxic atmospheres.

(2) Section 4.3 states that the procedures for testing the atmosphere and for entry should be decided on the basis of the preliminary assessment. In this section, risk is divided into three categories: Minimal risk to health or life, no immediate risk to health or life, and a risk to health or life.

(3) Sections 4.4 and 4.5 state that when the preliminary assessment indicates minimal risk to health or life, the precautions described in Sections 5, 6, 7, and 8 should be followed. Additionally, when the preliminary assessment indicates a risk to life or health, the additional precautions described in Section 9 should also be followed. Section 5 describes authorization of entry; Section 6 describes general precautions; Section 7 describes testing the atmosphere; Section 8 describes precautions during entry; and Section 9 describes additional precautions for entry into a space where the atmosphere is known or suspected to be unsafe.

(4) Section 8.1 states that the atmosphere should be tested at regular intervals while the space is occupied and people should be instructed to leave the space if conditions deteriorate.

Fatality of chemical tanker *KYOKUHO-MARU No. 2*

The fatality of a crewmember of the *KYOKUHO-MARU No. 2*, Japan flagged 388 gross tons chemical tanker, is a sample of how to find a hole in an actual case (JTSB 2011a).

The accident occurred at Kawasaki Ku of Keihin Port at about 13:55 on 10 March 2010. The C/O was suffocated by a lack of oxygen in a cargo tank during an unloading operation. The ship owner and general safety manager did not prescribe in a manual that access hatches should not be opened during a cargo operation because they thought such information was common knowledge among crewmembers onboard chemical tankers.

The *KYOKUHOU-MARU No. 2* had unloaded cargos, including tertiary butyl alcohol (TBA), nine times at the same berth in Kawasaki Ku of Keihin Port since January 2010. The general safety manager did not know that nitrogen gas was being charged into the cargo tanks during the unloading operation of TBA. The oxygen level in the cargo tank measured after the accident, at about 14:23, was 16 percent, while the acceptable oxygen level is 21 percent. The ship owner and general safety manager could not foresee that a crewmember would enter into the cargo tank without measuring the atmosphere and did not establish a procedure for enclosed space entry. This indicated that a hole in a defensive layer caused by latent conditions opened on the plan on the date that the manual was created and moved through the PDCA cycle embedded in the SMS until the accident occurred.

At the previous berth, the C/O ordered TBA to be loaded without putting a drain plug on the cargo piping which was removed for inspection prior to loading. The *KYOKUHOU-MARU No. 2* finished loading and proceeded to the berth for unloading, where the accident occurred. During unloading, the cargo was not unloaded. The C/O looked inside the cargo tank through the access hatch and noticed that the drain plug was not fixed on the cargo piping. He entered the cargo tank wearing a gas mask without measuring the atmosphere. When handling TBA, wearing PPE such as a gas mask was required (Figs. 8.8, 8.9).

At the local workplace, the procedure of unloading which did not establish the procedure of enclosed space entry was executed. This indicated that the hole in the defensive layer was created at the organization and moved to another defensive layer at the local workplace and the cargo

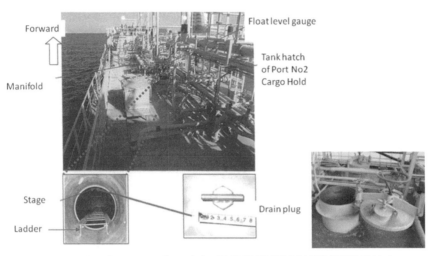

Fig. 8.8: Accident site onboard the *KYOKUHOU-MARU* (JTSB 2011a).

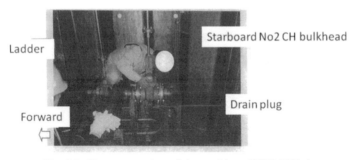

Fig. 8.9: Reconstruction of the accident (JTSB 2011a).

operation was conducted while the hole was opened until the accident occurred. This also indicated that the hole had been moving between the PDCA cycle at the organization and risk management process at the local workplace from the date that the manual was created to 10 March in 2010 because the procedure was approved or produced by the organization.

With regard to the *KYOKUHOU-MARU No. 2*, two defensive layers and their associated holes were related to the accident (Figs. 8.10, 8.11).

The unsafe act performed by the C/O was that, during cargo operation, he wore a gas mask and entered the cargo tank under conditions that the procedure for enclosed space entry had not been established.

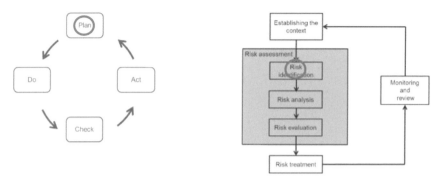

Fig. 8.10: Location of holes of the company (left) and the C/O.

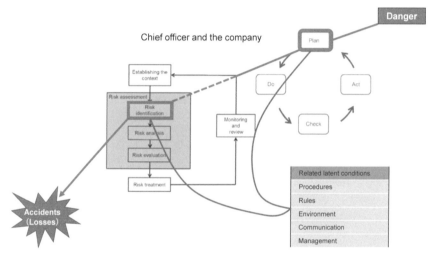

Fig. 8.11: Accident model on the *KYOKUHOU-MARU No. 2.*

These are latent conditions on the C/O

(1) The procedures did not prescribe that the access hatch should not be opened during loading and unloading operations.

(2) According to Ordinance of Industrial Safety and Health, ship owners were required to measure the oxygen concentration in cargo tanks before starting work in enclosed spaces or places where a risk of oxygen deficiency exists. These items were not carried out on the ship.

(3) In the "environment", the oxygen concentration in the cargo tank where the accident occurred at around 14:23 was 16 percent, hence, it was in a state of oxygen deficiency.

(4) The *KYOKUHOU-MARU No. 2* had unloaded TBA nine times at the same pier. The crewmembers, except for the C/O and the C/E, did not know that nitrogen gas had been injected into the cargo tank during the loading operation in order to prevent an explosion and to prevent negative pressure in the cargo tank. Therefore, critical safety information was not shared among crewmembers, and communication was incomplete.

(5) The ship owner did not take measures such as establishing a procedure manual of loading and unloading operations and implementing education and training on safety measures during TBA unloading operations; therefore, management was inappropriate.

These are preventive measures to be considered by the method of risk reduction:

(1) The ship owner should change the design of the access hatches to ensure that they do not open when the atmosphere is not measured (inherently safe design).

(2) The ship owner should establish a procedure for ensuring safety by means of measuring the atmosphere before entering a cargo tank, and provide education and training on the procedure for crewmembers. The ship owner should establish a system for sharing information with crewmembers in order to discern how effective the education and training has been.

(3) The master should put signs on access hatches indicating that access hatches should not be opened during cargo operations.

(4) Crewmembers should hold a risk assessment meeting on safety associated with cargo operations and share information before starting work.

(5) More than one crewmember should be present at inspection prior to loading and ensure that drain plugs are put in place properly.

(6) A consignee should share information with crewmembers through written and oral communications prior to cargo operations regarding hazardous substances charged into cargo tanks.

8.5.3 Grounding

Grounding is related to passage planning. Passage planning consists of appraisal, planning, execution, and monitoring (Swift 2000, IMO 2000b). It is considered that the appraisal stage covered risk identification; the planning stage covered risk analysis and risk evaluation; the execution stage covered risk treatment; and the monitoring stage covered monitoring and review.

A summary of passage planning is as follows:

(1) During the appraisal stage, the risks involved in the contemplated passage should be examined by gathering all relevant information, such as navigational charts, sailing directions, climatic information, draft of the ship, and personal experience.

(2) During the planning stage, a passage plan should be prepared on the basis of the appraisal stage. This plan should cover the entire voyage from berth to berth. When it becomes necessary to approach an area of potential danger, there are several essential rules that should be followed. The ship should always remain in safe water, sufficiently distant from any danger to minimize the possibility of grounding in the event of a machinery breakdown or navigational error. This part of the planning stage is considered as risk analysis. Risk evaluation is as follows:

A) No-go areas, where the ship cannot travel as a result of the relationship between the ship's draft and charted depths, should be marked on charts.

B) Safe water, the limits of which are bounded by margins of safety around the no-go areas, should be identified.

C) Tracks should be marked in the areas of safe water on the charts.

D) In tidal areas, time periods during which it is safe for the ship to travel with sufficient clearance should be shown.

E) In addition to tracks, abort positions where the ship cannot return to, as well as contingency plans, should be shown on charts.

(3) During the execution stage, the voyage should be executed in accordance with the passage plan.

(4) During the monitoring stage, the progress of the ship should be closely and continuously monitored in order to discern whether the ship is proceeding in accordance with the passage plan. Any changes to the passage plan should be consistent with these stages and clearly marked and recorded.

Grounding of the cargo vessel *LANA*

The grounding of the cargo vessel *LANA*, Cambodia flagged 1,272 gross tons cargo vessel, is a sample of how to find a hole in an actual case (JTSB 2011c).

The accident occurred in the northern waters of Mikomoto Shima (island) at about 20:39 on 15 December 2009 (Fig. 8.12). The master selected a planned route in the northern waters of Mikomoto Shima, which was not recommended by sailing directions, when he drafted a passage planning from Chiba port in Japan to Masan port in Republic of Korea. The *LANA* left Chiba port at about 10:00 the same day. The master made an improper risk analysis during the process of passage planning.

He saw the ship's position plotted by an able seaman on watch at 19:30 near her planned course and thought she was proceeding on the planned route. The master did not fix her position by himself during the process of monitoring of the passage planning while underway in the northern waters of Mikomoto Shima. This indicated that two holes in a defensive layer, caused by latent conditions, opened no later than about 10:00 and 19:30 on 15 December and moved through the process of passage planning until the accident occurred at about 20:39 the same day (Figs. 8.13, 8.14).

The unsafe act performed by the master was that he kept the ship's course and speeds without being aware of its proximity to Yokone (rock) in the no-go areas.

These are latent conditions on the master.

(1) The master did not adopt the route recommended by sailing directions; as a result, the planning stage of passage planning was not appropriate. In addition, the master did not fix the ship's position by himself, the monitoring stage of the passage planning was not carried out.

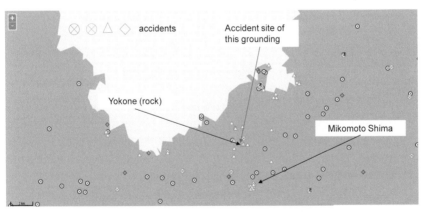

Fig. 8.12: Accident site and accident data from 19:85 to 20:17 (JTSB).

Color version at the end of the book

177

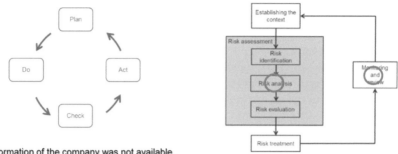

Information of the company was not available.

Fig. 8.13: Location of holes of the master of the *LANA*.

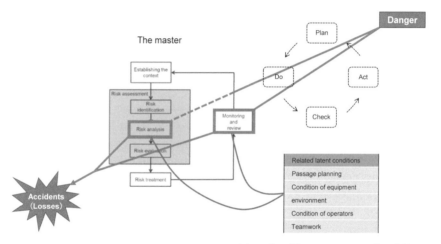

Information of the company was not available.

Fig. 8.14: Accident model on the *LANA*.

(2) The *LANA* was equipped with two radars, but one failed and the other was not connected to the gyro compass. Under these conditions of radar, the master could not use a parallel index to find ship's position.

(3) In the "environment", there were many rocks between Mikomoto Shima and the south coast of the Izu Peninsula.

(4) The master had never experienced sailing off northern waters of Mikomoto Shima during the night. This case was the first time for the master to navigate the channel.

(5) The master left fixing ship's position to the able seaman who did not and was not required to learn about fixing positioning, and the master did not fulfill his responsibility.

These are preventive measures to be considered by the method of risk reduction:

(1) The master should select a planned route through the southern waters of Mikomoto Shima, which is recommended by sailing directions (inherently safe design).

(2) The officer on the watch should monitor the ship's progress by position fixing in order to determine whether the vessel is proceeding in accordance with passage planning.

8.5.4 Occupational casualties

In this case, procedures for lowering a rescue boat are based on the manual and checklists in accordance with the SOLAS Convention aboard the *ANNA MAERSK,* Denmark flagged 93,496 gross tons container vessel (JTSB 2014d).

The lowering procedure of the rescue boat was as follows: (1) Confirm that the rescue boat is connected to the hook, (2) Hoist the rescue boat from the cradle by winding up the wire rope with the crane winch, and then swing it outside of the ship by turning the crane boom, (3) Lower the rescue boat to the same level of the deck by extending the wire rope, (4) Able seaman is to board, (5) C/O is to board, (6) Lower the rescue boat onto the sea surface by extending the wire rope, (7) Able seaman is to release the suspension from the rescue boat.

Fatalities of crewmembers of the *ANNA MAERSK*

The accident occurred at Kobe Ku of Hanshin Port during a drill of lowering a rescue boat at about 11:15 on 27 March 2012. The C/O suffered a serious injury and an able seaman died. The accident investigation completed in January 2014 found that the shackle pin of the swivel included in the off-load release hook device of a rescue boat onboard the *ANNA MAERSK* did not have safety measures against breakaway (Figs. 8.15, 8.16, 8.17).

Concerning the safety of the off-load release hook device, the device passed a load test conducted by a shipyard on 25 March 2008. The cargo ship safety equipment certificate was issued by a classification society on 6 January 2011. No comment on the device was made by the internal

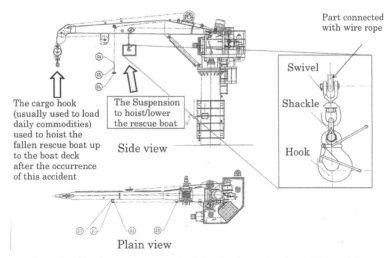

The cargo hook (usually used to load daily commodities) used to hoist the fallen rescue boat up to the boat deck after the occurrence of this accident

The Suspension to hoist/lower the rescue boat

Side view

Part connected with wire rope

Swivel

Shackle

Hook

Plain view

Fig. 8.15: Design of crane and off-load release hook (JTSB 2014d).

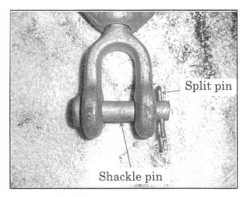

Split pin

Shackle pin

Fig. 8.16: Shackle pin and split pin (JTSB 2014d).

audit conducted by the ship management company on 31 December 2011. Risk assessment on the safety of lowering the rescue boat was conducted by a crewmember without considering safety measures to be taken if the device broke down in 2011.

The crewmember did not check the condition of the swivel before the accident occurred. The C/O was following the procedure of lowering the rescue boat at the time of the accident. Persons at the organization and the local workplace could not have foreseen the breakdown of the off-load release hook device during the drill.

180

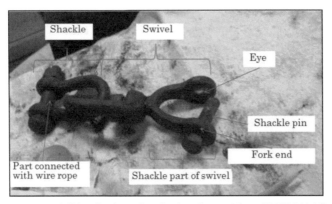

Fig. 8.17: Off-load release hook after the accident (JTSB 2014d).

This indicated that a hole in a defensive layer caused by latent conditions opened during the process of plan and moved through the PDCA cycle embedded in the Safety Management System until the accident occurred. Because the interval period between drills was regulated by the SOLAS Convention, the hole in the defensive layer at the organization moved to another defensive layer at the local workplace and the hole had been moving through risk management process embedded in the procedure at the local workplace until the accident occurred. This also indicated that the hole had been moving between the PDCA cycle at the organization and risk management process at the local workplace from the date that the off-load release hook device was rigged to 27 March 2012 because the procedure was approved or produced by the organization (Fig. 8.18).

With regard to the *ANNA MAERSK*, two defensive layers and their associated holes were related to the accident. The C/O was following the procedure of lowering the rescue boat at the time of the accident; therefore, no unsafe act was performed (Fig. 8.19).

These are latent conditions on the vessel.

(1) Prevention of detachment of the shackle pin of off-load release hook was totally dependent on a split pin attached to the shackle pin. Safety measures against the breakaway of the split pin had not been taken into consideration for prevention of detachment of the shackle pin due to shear fracture of the split pin. This was a design issue.

(2) The ship management company did not take safety measures to prevent the detachment of the shackle pin, and the management was inappropriate.

These are preventive measures to be considered by the method of risk reduction.

(1) The ship owner should change the design of the swivel to be composed of a bolt, nut and split pin (inherently safe design).

(2) The ship management company should designate the off-load release hook device to be checked, and make crewmembers conduct visual checks on items.

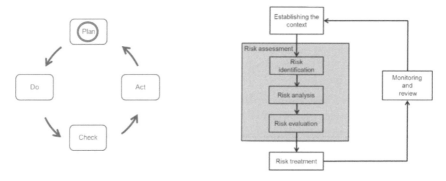

No hole opened in the local workplace.

Fig. 8.18: Location of holes of the company (left) and the C/O.

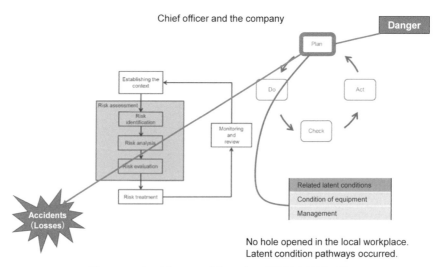

No hole opened in the local workplace.
Latent condition pathways occurred.

Fig. 8.19: Accident model on the *ANNA MAERSK*.

8.6 Holes and latent conditions

It appears that each type of accident has different contributing factors, considering Chapter 4. If the relationship between holes and latent conditions becomes clear, it is possible to identify frequent contributing factors and also determine the pattern of each accident.

Eighty-nine serious marine accidents that were investigated and published by JTSB from 2008 to 2014 were analysed by using the methodology shown from Sections 8.3 to 8.5 (Fukuoka 2016b, Fukuoka and Furusho 2016a).

8.6.1 Selection of samples

This study does not address marine incidents, which are events other than a marine accident that would endanger the safety of the vessel, people or the environment, if not corrected (IMO 2008). An incident differs from an accident with regard to the causal mechanism. In an accident, the defenses do not work, and loss or damage ensues, whereas in an incident, the defenses work (ICAO 1993, Hollnagel 2004). The IMO and the ICAO both use the same definition for an incident and an accident.

From 2008 to 2014, there were 6,974 marine accidents and 1,089 marine incidents involving 10,759 vessels, of which 5,083 were classified as other than fishing vessels, recreational fishing vessels, angler tender boats, pleasure boats, or Personal Water Craft (PWC). Approximately 10 percent of these 5,083 vessels, i.e., 508 vessels, were involved in marine incidents, according to the statistics (JTSB 2012c, 2013, 2014f, 2015b). Therefore, the remaining 4,575 vessels were considered to have been involved in organizational accidents. The 89 cases included in this study are regarded as a representative sample of these organizational accidents.

The type of accidents is divided into 8: Collision, contact, grounding, occupational casualties, fires, explosions, sinking and capsizing. Contact means an event where a ship strikes a breakwater, quay, pier, light buoy, or other facilities, causing damage to a ship or facility. Occupational casualty means an event in which crew members, passengers, workers, etc., are killed or missing in connection with the structure, equipment and operation of ships. Occupational casualty often happens during loading and unloading operations and drills of rescue boats and life boats.

8.6.2 Selection of local workplaces and organizations

A local workplace is characterized as a vessel that was involved in a marine accident and an organization is, in most cases, characterized as a ship management company. In the case of a collision, two vessels are involved; hence, a collision is considered to involve two local workplaces. Other types of marine accidents involve one vessel, i.e., one local workplace.

When accidents occurred in organizations other than ship management companies, such as pilot associations, equipment manufacturers and cargo-handling companies, these organizations were also studied in terms of holes and latent conditions. When there was a pilot or berth master on board a vessel, there were two defensive layers of risk management at that local workplace: The pilot or berth master and the master of the vessel. Holes and latent conditions were studied in relation to each pilot or berth master and vessel master.

When the marine accident investigation reports did not contain any local workplace factors or organizational factors, the study could not analyze any holes or latent conditions.

8.6.3 Number of defensive layers in a local workplace (vessel)

Five vessels or local workplaces had four defensive layers; four vessels were operated by the pilots or berth master who was on board in cases involving collisions or contacts, while the other two vessels were associated with an equipment manufacturer or a cargo-handling company, in addition to a ship management company, in cases involving occupational casualties.

Three vessels had three defensive layers: One vessel was operated by the pilot in a case involving a collision, but a pilot association was not included by the marine accident investigation report, while the two other vessels were associated with a ship management company and a cargo-handling company in cases involving occupational casualties, but local workplace factors on the part of crew members of the vessel were not included by the marine accident investigation report.

Forty-four vessels had two defensive layers: An SMS and risk management. Fifty-seven vessels had one defensive layer, namely risk management; they did not have an SMS because the marine accident investigation reports did not include organizational factors.

Ten vessels did not have a defensive layer; nine of these were fishing vessels, and local workplace factors were not included, owing to the death of the captain after collisions. The other vessel was a passenger vessel, for which organizational factors were not included and there was no active failure on the part of an operator (Table 8.2).

8.6.4 Locations of holes in organizations

Sixty-five organizations were analyzed in this study: 28 cases involved collisions, five contact, two grounding, 24 occupational casualties, three fire, one explosion, and one involved sinking. In this category, sinking was not preceded by any other accident.

In eight of these organizations, the SMS defensive layer had two holes. These eight organizations had another defensive layer of risk management at their local workplaces. In the other organizations, the SMS defensive layer had one hole.

The most frequent hole in the SMS defensive layer opened during the "plan" process of the PDCA cycle and accounted for 45 percent of the total number of holes. The second most frequent hole opened during the "do" process and accounted for 44 percent of the total number of holes. Therefore, holes that opened during the "plan" and "do" processes accounted for 89 percent of the total number of holes (Table 8.3).

In one case involving sinking, a hole opened during the "check" process, indicating that procedures in relation to audits and reporting accidents and non-conformities were not carried out. These findings indicate that holes in the SMS defensive layer tend to arise during the early stages of the PDCA cycle, except for cases involving sinking.

8.6.5 Locations of holes at local workplaces

The total number of local workplaces analyzed in this study was 121: 70 cases involved collisions, seven contact, nine grounding, 25 occupational casualties, four fire, two explosion, two sinking, and two involved capsizing.

The most frequent hole in the defensive layer of risk management opened during risk analysis and accounted for 49 percent of the total number of holes. The second-most frequent hole opened during risk identification and accounted for 31 percent of the total number of holes.

Table 8.2: Number of defensive layers.

	Collisions	Contacts	Groundings	Occupational casualties	Fires or explosions	Sinkings or capsizings	Total
Number of serious marine accidents	35	7	9	26	7	5	89
Number of local workplaces (vessels)	70	7	9	26	7	5	124
Four defensive layers	3	1	0	1	0	0	5
Three defensive layers	1	0	0	2	0	0	3
Two defensive layers	23	3	3	15	2	0	46
One defensive layer	35	3	6	8	4	5	61
Zero defensive layers	8	0	0	0	1	0	9

Table 8.3: Locations of holes observed at 65 organizations and 121 local workplaces in 89 cases of serious marine accidents.

Locations	Collisions	Contact	Grounding	Occupational casualties	Fires or explosions	Sinking or capsizing	Total
Organization							
Plan	12	1	1	17	2	0	33
Do	13	3	2	12	2	0	32
Check	4	1	1	1	0	1	8
Act	0	0	0	0	0	0	0
Total number of holes in organizations by the types of accidents	29	5	4	30	4	1	73
Local workplaces							
Risk identification	12	0	3	17	3	3	38
Risk analysis	48	0	4	7	1	1	61
Risk evaluation	1	1	1	0	0	0	3
Risk treatment	6	2	2	0	0	0	10
Monitoring and review	0	6	6	0	0	0	12
Total number of holes at local workplaces by the types of accidents	67	9	16	24	4	4	124

187

Therefore, the holes that opened during risk analysis and risk identification accounted for 80 percent of the total number of holes. The most frequent hole in the defensive layer of risk management was not necessarily the same for different types of accidents (Table 8.3).

In cases involving collisions, the most frequent hole opened during risk analysis; in cases involving contact and groundings, it opened during the monitoring and review process; in cases involving occupational casualties, fire or explosion, it opened during risk identification. These findings indicate that holes in the defensive layer of risk management tend to arise during the early stages of the risk management process in cases involving collisions, occupational casualties, fire, or explosion, and that they tend to arise later during the process in cases involving contacts. In cases involving grounding, they arise during all stages in the process, but mostly during the monitoring and review stage.

8.6.6 Latent conditions

A total of 518 latent conditions were analyzed and categorized into 10 groups. Condition of operators was the most frequent latent condition that caused holes to open and accounted for 19 percent of the total number of latent conditions (Table 8.4, Fig. 8.20). Psychological limitations were the most frequent subdivision in condition of operators (Table 8.5). Environment was the second most frequent latent condition and accounted for 16 percent of the total number of latent conditions. Traffic density was the most frequent subdivision in environment (Table 8.6).

The most frequent latent conditions were not necessarily the same for different types of accidents. In some cases, multiple subdivisions of condition of operators and environment were selected. With regard to fires or explosions, the latent conditions were almost the same, but the accident sites on the vessels differed. Fires occurred in the cargo hold of a car carrier, an accommodation room, and engine rooms, while explosions occurred in enclosed spaces such as a cargo hold.

8.7 Abstract generalizations (general characteristics of accidents)

With regard to latent conditions, condition of operators was the most frequent latent condition that caused holes to open. However, the most frequent latent conditions were not necessarily the same for different

Table 8.4: Number of latent conditions in 89 cases of serious marine accidents.

Latent conditions	Collisions	Contact	Grounding	Occupational casualties	Fires or explosions	Sinking or capsizing	Total
Passage planning	15	8	9	4	0	3	39
Procedures	13	1	0	26	7	2	49
Rules	64	0	1	4	2	0	71
Human machine interface	7	0	0	1	0	0	8
Condition of equipment	3	1	4	7	4	3	22
Environment	46	7	10	15	3	4	85
Condition of operators	64	6	8	13	4	2	97
Communication	28	3	7	7	1	1	47
Team work	19	2	4	10	0	0	35
Management	28	5	3	24	4	1	65
Total number of latent conditions by the types of accidents	287	33	46	111	25	16	518

Table 8.5: Subdivisions of condition of operators.

Subdivisions of condition of operators	Collisions	Contact	Grounding	Occupational casualties	Fires or explosions	Sinking or capsizing	Total
Physical limitations	0	0	0	0	0	0	0
Physiological conditions	3	1	1	2	0	0	7
Psychological limitations	60	5	7	4	3	1	80
Individual workload management	0	0	0	0	0	0	0
Knowledge, skills, experience, education, and training	5	2	3	10	2	1	23
Total	68	8	11	16	5	2	110

Table 8.6: Subdivisions of environment.

Subdivisions of environment	Collisions	Contact	Grounding	Occupational casualties	Fires or explosions	Sinking or capsizing	Total
Sea and weather conditions	7	1	2	4	0	4	18
Conditions in which people are working, such as atmosphere in cargo hold	0	0	0	7	3	0	10
Traffic density	25	4	0	0	0	0	29
Geographical features, such as a narrow channel	17	1	9	0	0	0	27
Berth facilities and other factors	0	2	0	4	0	0	6
Total	49	8	11	15	3	4	90

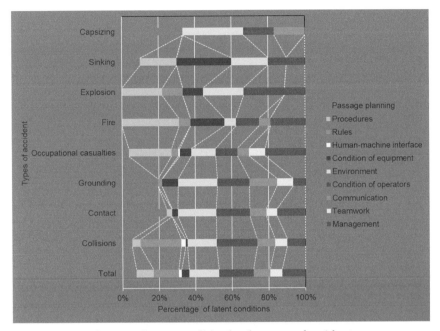

Fig. 8.20: Latent condition by the types of accident.

types of accidents. Here, frequent latent conditions are defined as those that account for 80 percent of the total number of latent conditions for a particular type of accident. The relationship between latent conditions and the characteristics of holes and an abstract generalization about the situation in which holes tend to arise in relation to different types of marine accidents are presented in the following sections.

Abstract generalizations drawn from the results of this study are compatible with the common patterns of causality for both collisions and grounding examined by Macrae (2009). Macrae concludes, after analyzing 30 marine accidents investigation reports published by the Australian Transport Safety Bureau, that collisions often involved a fishing vessel and a cargo vessel and resulted from a problem identifying the existence or speed of the other vessel, while grounding resulted from inadequate passage plan, coupled with either a problem locating the vessel or communication problems on the bridge.

In this book, the number of cases of fire, explosion, sinking and capsizing are small; therefore, abstract generalizations are limited for collisions, contact, groundings and occupational casualties.

8.7.1 Cases involving collisions

The most frequent hole opened during risk analysis, accounting for 72 percent of the total number of holes. This result indicated that, when following the procedures to avoid a collision, most vessels did not use all of the available means to determine whether any collision risk existed; radar equipment was not used properly, and assumptions were made on the basis of insufficient information. In organizations, the most frequent holes opened during the "plan" and "do" processes in the PDCA cycle, and these accounted for 86 percent of the total number of holes.

Rules, condition of operators, environment, communication, and management were the most frequent latent conditions, and accounted for 80 percent of the total number of latent conditions (Fig. 8.21). Psychological limitations accounted for 88 percent of the total number of subdivisions in condition of operators. Traffic density and geographical features such as a narrow channel accounted for 86 percent of the total number of subdivisions in environment. With regard to collisions, the author's previous study shows that assumptions accounted for 82 percent of psychological limitations (Fukuoka 2015).

These findings indicate that holes tend to open when operators make assumptions during risk analysis regarding the risk of collisions while vessels are navigating in congested waters or waters that have geographical features such as narrow channels, and that operators lack

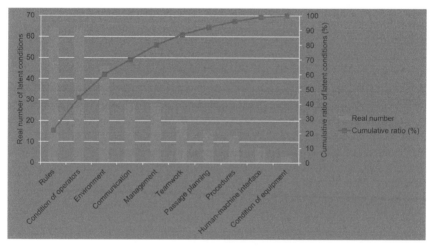

Fig. 8.21: Pareto chart on collisions.

communication with the bridge team or other people on safety issues. Procedures regarding the safe operation of ships in these types of waters are either not established by organizations or, where established, are not followed by operators.

8.7.2 Cases involving contact

The most frequent hole opened during monitoring and review, and these accounted for 67 percent of the total number of holes. This result suggested that during the monitoring stage of passage planning, ship's progress was not closely and continuously monitored. In organizations, the most frequent hole opened during the "do" process.

Passage planning, environment, condition of operators, and management were the most frequent latent conditions, and accounted for 79 percent of the total number of latent conditions. Psychological limitations accounted for 63 percent of the total number of subdivisions in condition of operators. Traffic density accounted for 50 percent of the total number of subdivisions in adverse environment (Fig. 8.22).

These findings indicate that holes tend to open when operators' psychological limitations become evident during the monitoring and review of the ship's progress while vessels are navigating in congested

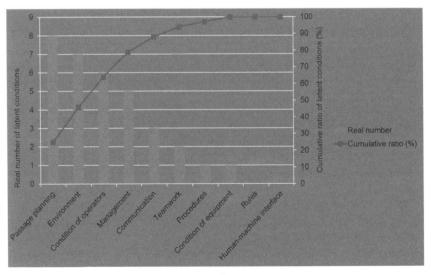

Fig. 8.22: Pareto chart on contact.

193

waters. Procedures on the safe operation of ships in these waters that are established by organizations are not followed by operators.

8.7.3 Cases involving grounding

The most frequent hole opened during monitoring and review, and the second most frequent hole opened during risk analysis. These holes accounted for 64 percent of the total number of holes. This result indicates that the ship's progress was not closely and continuously monitored. Moreover, during the passage planning stage, the master did not observe the essential rule that a ship should always remain in safe water, sufficiently distant from any danger to minimize the possibility of grounding in the event of a machinery breakdown or navigational error. In organizations, the most frequent hole opened during the "do" process.

Environment, passage planning, condition of operators and communication were the most frequent latent conditions, and accounted for 74 percent of the total number of latent conditions. Psychological limitations accounted for 64 percent of the total number of subdivisions in condition of operators. Traffic density and geographical features such as a narrow channel accounted for 82 percent of the total number of subdivisions in adverse environment (Fig. 8.23).

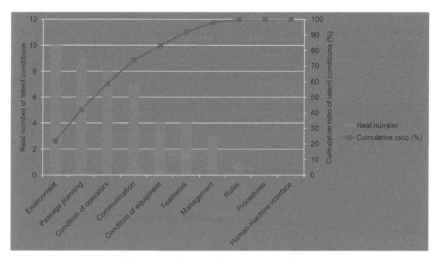

Fig. 8.23: Pareto chart on grounding.

These findings indicate that holes tend to arise when operators are subject to psychological limitations during monitoring and review of ship's progress while vessels are navigating in congested waters or waters with geographical features such as narrow channels, and that operators do not engage sufficient communication with the bridge team or other people on safety issues. During the passage planning stage, the master does not observe the minimum rules for the safe operation of the ship. Procedures established by the organizations regarding the safe operation of ships are not followed by operators in these waters.

8.7.4 Cases involving occupational casualties

The most frequent hole opened during risk identification, and accounted for 72 percent of the total number of holes. This result suggested that during work, such as loading and unloading cargo or entering cargo tanks, most operators did not identify hazards in local workplaces. In organizations, the most frequent holes opened during the "plan" and "do" processes, and these accounted for 93 percent of the total number of holes.

Procedures, management, environment, condition of operators and teamwork were the most frequent latent conditions, and accounted for 79 percent of the total number of latent conditions. Knowledge, skill, experience, education, and training accounted for 63 percent of the total number of subdivisions in condition of operators. Conditions in which people were working, such as the atmosphere in cargo tanks, accounted for 47 percent of the total number of subdivisions in environment (Fig. 8.24).

These findings indicate that holes tend to arise when operators work in variable conditions, for instance, in cargo tanks where the atmosphere can deteriorate, and procedures relating to entering the cargo tanks are either not established by the organizations or, where established, are not used by the operators. There is inadequate teamwork and the operators are unable to identify hazards in local workplaces, mainly because of a lack of appropriate knowledge, skill, experience, education and training.

8.8 Unresolved issues on the SCM

These are issues that have been disputed on the SCM, and that are solved by a recent study (Fukuoka 2016b).

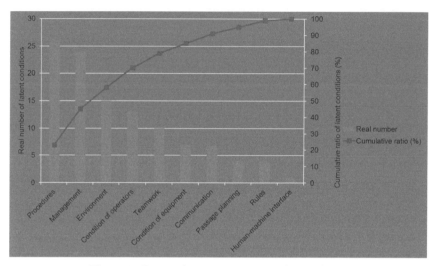

Fig. 8.24: Pareto chart on occupational casualties.

(1) What constitutes a hole.

Regarding what constitutes a hole, it is defined in this chapter that a hole arises when the risk is in an "unacceptable region" that indicates that the risk is not justified for whatever reason. In an organization, an unacceptable risk situation is defined as one in which the functional requirements related to accident prevention, as prescribed by the ISM Code which is based on quality management systems of International Standards, were not satisfied. Content of the hole in the organization can be explained by what constitutes each process of the PDCA cycle in Chapter 5, and its example is shown in Section 5.6.

At a local workplace, the concrete definition of a hole is shown by each type of accident as follows. In collisions, a situation in which two vessels approach within the maximum advance of a give-way vessel or within tactical diameter if they collide while overtaking. In contacts, a situation in which a vessel approaches such danger as breakwaters or aquaculture facilities within tactical diameter. In groundings, a situation in which a vessel enters a no-go area and encounters dangers such as a rock, a sunken rock or shallow waters. In occupational casualties, a situation in which, for instance, a person is about to enter an enclosed space during loading or unloading operation and the person does not measure the atmosphere in the enclosed space prior to entering it. Since occupational casualties involve various kind of accidents, the concrete definition of a hole requires that each organization agree and decide where the unacceptable region will be for each kind of accident. In a fire, a situation in which combustible materials such as fuel oil or lubricating oil leak from the container which holds them where an ignition source is present. In an explosion, a situation in which a person uses flames without detecting the atmosphere where flammable gases are present. In sinking or capsizing, a situation in which the ship's righting moment decreases, and its stability is at risk. Since stability is closely related to GM, which varies with ship type and each vessel, organizations and operators need to constantly recognize the situation of GM.

(2) Why a hole opens.

The reason that a hole opens in the SCM was shown by clarifying the relationship between a hole and the latent conditions in each type of accident. Latent conditions that cause a hole to open are divided into 10 categories that can be applied to marine accidents in this chapter. Based on statistics from marine accident investigation reports issued by JTSB, abstract generalizations or why holes open and an accident occurs, were revealed by each type of accident. In collisions, for instance, a hole opens when two vessels approach within the maximum advance of a give-way vessel. And this situation appears mostly when operators make assumptions during risk analysis regarding the risk of collisions while vessels are navigating in congested waters or waters that have geographical features such as narrow channels, and that operators lack communication with the bridge team or other people on safety issues. Procedures regarding the safe operation of ships in these types of waters are either not established by organizations or, where established, are not followed by operators.

(3) Location and movement of a hole.

The location and movement of a hole was clarified by demonstrating the following contention that Reason made. Reason (1990) states that no one can foresee all possible accident scenarios and, as a result, some holes in defensive layers caused by latent conditions will be present from the time of system establishment or will develop unnoticed or uncorrected during system operation. The results of the study in this chapter show that the location of the hole depends on the type of accident and is not necessarily the same, and the hole was confirmed to move on the SMS defensive layer and on the risk management defensive layer.

(4) Opening and closing of a hole

Regarding the opening and closing of a hole, the concept of risk was applied. In this chapter, the opening of a hole in a defensive layer is defined as an unacceptable risk in an organization or local workplace. An experimental study was conducted on four actual marine accidents, each process of SMS defensive layer and risk management defensive layer was focused on and it was found that holes opened in the process of the defensive layers. Regarding the closing of the holes, it was also confirmed that if the risk reduction method of ISO/IEC Guide 51 is used, it is possible to take a measure to close the holes in the defensive layer in the above four cases of marine accidents.

(5) Why holes need to line up to cause an accident

As described in this chapter, out of 89 marine accidents in this study, there were 13 cases where two or more holes opened in one defensive layer. It became clear that the holes could not line up, so it was recognized that the accident trajectory was not straight. The result is different from the theory of the SCM, that accidents occur when the holes in defensive layers line up and the accident trajectory goes through all the holes.

8.9 The SCM in maritime industry

8.9.1 Application of the SCM to the marine accidents

It was found that four explanations of the SCM are applicable to the marine accidents (Fukuoka 2016b, Fukuoka and Furusho 2016a).

(1) The findings support the existence of in depth defenses between danger and accidents. In this study, 54 local workplaces or vessels had two defensive layers at the time of the accident. When a pilot or berth master was on board, a local workplace had four defensive layers. Sixty-one local workplaces or vessels had only one defensive layer, that of risk management, because organizational factors were not included in the marine accident investigation reports. If organizational factors are investigated and included in marine accident investigation reports, it is considered that a local workplace or vessel has at least two defensive layers, the PDCA cycle and risk management, at the time of the accident.

(2) The findings support the dynamic movement of the holes. Table 8.3 indicates that the locations of holes that opened differed. However, trends were evident in the locations of holes. In organizations, except for one case involving sinking, the holes tended to open early in the PDCA cycle. In local workplaces, except for cases involving contact and grounding, they tended to open early in the risk management process. In cases involving contact, they tended to open later in the process, while in cases involving grounding, they arose at all stages of the process.

(3) The findings mentioned above (2) support Reason's statement that no one can foresee all possible accident scenarios, and that some holes in defensive layers will be present right from the time of system establishment or will develop in an unnoticed or uncorrected manner during system operation.

(4) The findings support the latent condition pathways. Reason (1997) states that unsafe acts are not a necessary condition for organizational accidents, and that on some occasions the defenses fail simply as a result of latent conditions, as in the Challenger Space Shuttle and King's Cross Underground fire disasters. He refers to these situations as latent condition pathways. In three cases involving occupational casualties, two cases involving fire, and one case involving sinking, the defensive layer of risk management did not have any holes because there were no active failures on the part of operators. In this study, latent condition pathways were seen to occur owing to the condition of equipment.

8.9.2 Shortcomings of the application of the SCM

It was found that three explanations of the SCM are not applicable (Fukuoka 2016b, Fukuoka and Furusho 2016a).

(1) There is, on some occasions, more than one hole in a defensive layer of the SCM in the real world. In this study, the manner in which holes in defensive layers lined up to enable an accident to occur could not be clarified. In eight organizations, the SMS defensive layer had two holes. At five local workplaces, the defensive layer of risk management had two or more holes.

(2) The findings suggest that the accident trajectory is not necessarily straight. According to the SCM, the accident trajectory passes through

all the aligned holes. However, if one defensive layer has two holes, the accident trajectory cannot pass through both holes. The study indicates that an accident occurs when the accident trajectory passes through all the defensive layers. Therefore, if an SMS and risk management are defensive layers in the SCM, this suggests that an accident trajectory that passes through all holes is not straight.

(3) All the holes in a defensive layer need to be closed to prevent an accident from occurring. According to the SCM, when the holes in defensive layers are lined up, the accident trajectory passes through all the holes, and the accident occurs. In other words, only one hole in each defensive layer is related to the accident, and the accident can be prevented when one hole in each defensive layer is closed. However, the findings indicate that all holes in a defensive layer need to be closed to prevent an accident because the accident trajectory is not straight.

(4) On the SCM, defensive layers are drawn in series. In the case of the collision between the car carrier *MEDEA* and the fishing vessel *KOSEIMARU*, defensive layers were not able to be arranged in series but in parallel. Therefore, there is a possibility that SCM on defensive layers did not assume an organizational accident caused by simultaneous presence of operators belonging to multiple organizations at local workplace. In studying 89 cases of serious marine accidents, there were 10 cases where defensive layers were drawn in parallel.

8.10 Conclusions

Making use of the concept of risk, risk management, the PDCA cycle and the SHEL model to clarify a hole, defensive layers and latent conditions on the Late 1 SCM, and putting them into a real marine accident, new knowledge about the unsolved issues on the SCM and the trend of occurrence of holes and latent condition were found. It was also found that the location of frequent holes and emergence of frequent latent conditions varied by types of accident; consequently, it explains how collisions, contact, grounding and occupational casualties are occurring.

Chapter 9

Statistical Science and Characteristics of Each Types of Accidents

9.1 Introduction

National marine accident investigation authorities conducted an analysis of causal factors by using the SCM as theoretical background after collecting evidence with the SHEL model in accordance with the Casualty Investigation Code (CIC) adopted by the IMO. The SCM posits that if latent conditions in a system are identified in advance it may be possible to take preventive measures before accidents take place.

Statistics can depict accurate trends and in some cases prediction is possible. This study quantified the degree of influence of each element of SHEL or latent conditions on the occurrence of accidents by conducting a multivariate analysis on statistical data attained from 89 cases of serious marine accidents, which is the same as indicated in the previous chapter (Fukuoka and Furusho 2016b). Additionally, features of frequent hole location on defensive layers in each type of accident were described by using correlation coefficient (Fukuoka 2016b).

9.2 Quantification of the SHEL element

9.2.1 Analytical method

The causal factors of serious marine accidents were classified according to each element of the SHEL model. Table 8.4 in the previous chapter shows the relationship between each element of SHEL and 10 latent conditions. The occurrence of accidents and each element of SHEL have a cause and effect relationship; therefore, the number of vessels involved in a serious marine accident is regarded as the dependent or y variable, and each element of the SHEL model as an independent or x variable.

In this study, x_1 represents software that addresses passage planning, procedures and rules; x_2 is hardware, which includes human-machine interfaces and condition of equipment; x_3 is environment; x_4 is central liveware or L1; and x_5 is peripheral liveware or L2, which includes communication, teamwork and management. Variable y represents the number of vessels involved in each type of accident. Variable n represents the number of each type of accident: Collisions, contacts, grounding, occupational casualties, fires, explosions, sinking and capsizing.

Multiple regression analysis was conducted on these statistical data using Excel *Toukei* 2015 for Windows.

9.2.2 Multiple regression equation

Table 9.1 shows statistics on the number of vessels involved in accidents (as dependent variables) and elements of the SHEL model (as independent variables). Table 9.2 shows the result of multiple regression analysis. The multiple regression equation, which was calculated using the data shown in Table 9.1, was as follows:

$$y = 0.428 + 0.514x_1 + 0.250x_2 + 0.051x_3 + 0.014x_4 + 0.125x_5$$

R^2 was 0.999 and adjusted R^2 was 0.999.

9.2.3 Hypothesis testing of multiple regression equation

The degrees of freedom were 5, and the multiple regression equation was significant at $p < 0.001$.

Table 9.1: The number of vessels and elements of the SHEL model.

Types of accident	y	S(X_1)	H(X_2)	E(X_3)	L1(X_4)	L2(X_5)
Collisions	63	92	10	46	64	75
Contacts	7	9	1	7	6	10
Grounding	9	10	4	10	8	14
Occupational casualties	26	34	8	15	13	41
Fires	5	6	3	1	2	4
Explosions	2	3	1	2	2	1
Sinking	3	3	3	2	1	1
Capsizing	2	2	0	2	1	1

Table 9.2: The result of multiple regression analysis.

Independent variables	Partial regression coefficient	Standard error	Standardized partial regression coefficient
X_1	0.514	0.196	0.755
X_2	0.250	0.140	0.042
X_3	0.051	0.325	0.036
X_4	0.014	0.301	0.015
X_5	0.125	0.172	0.158
Intercept	0.428	0.556	−

9.2.4 Influence of latent conditions on accidents

The influence of independent variable on dependent variable is measured by the standardized partial regression coefficient. The influential elements of the SHEL model, in descending order, were: Software ($\beta = 0.755$), peripheral liveware ($\beta = 0.158$), hardware ($\beta = 0.042$), environment ($\beta = 0.036$) and central liveware ($\beta = 0.015$).

The most influential element of the SHEL model on the number of vessels involved in accidents was software, and then peripheral liveware. The standardized partial regression coefficients of hardware, environment and central liveware were orders of magnitude smaller than those of software and peripheral liveware.

Chauvin et al. (2013) studied 39 vessels involved in cases of collisions and classified causal factors in accordance with the HFACS, which was based on the SCM. They concluded that decision errors and perceptual errors of operators accounted for 97 percent of operators' unsafe acts, and that the most frequent causal factors were communication with approaching vessels, BRM, safety management systems and audits. Although the research of Chauvin et al. dealt with the cases of collisions only and the classification system of causal factors differed from those of this study, the observed importance of communication, BRM, and safety management systems for collisions in their study meant that peripheral liveware defined in this study. Therefore, their results were consistent with this study, in that peripheral liveware had a strong influence on the number of vessels involved in accidents.

Despite the fact that the percentage of central liveware was 19 percent of the latent conditions of the overall marine accidents, the degree of influence on the number of accident occurrences was 1.5 percent. Swift (2000) and Adams (2006) also emphasize the importance of the condition of operators, such as distraction, stress, complacency and fatigue. The results of this study clearly show that condition of operators has the smallest degree of influence on the number of accident occurrences. In order to reduce the number of accidents, it is considered that even if improving the condition of operators, such as the psychological limitations, the influence is small.

9.3 Relationship between the hole location and the number of hole occurrence

The relationship between the hole location and the number of hole occurrences is a correlation without causality. Therefore, defining the hole location as x, and the number of holes occurrence as y, the relationship between these two variables is calculated by Pearson's correlation coefficient (hereinafter referred to as "correlation coefficient"), and the general tendency of the hole location is clarified by accident type.

9.3.1 Analytical method

For quantification in local workplaces, using Table 8.3 in the previous chapter, risk identification of the risk management process was defined as 1, risk analysis as 2, risk assessment as 3, risk treatment as 4 and monitoring and review as 5. In organizations, the plan of the PDCA cycle was defined as 1, do as 2, check as 3 and act as 4.

As a method of analyzing the tendency of the hole location by accident type, the correlation coefficient r, representing the strength of the correlation, was used.

It is assumed that the correlation strength is "very strong" positive or negative when the correlation coefficient r is ± 0.7 or more, "strong" when it is about ± 0.4, "weak" about ± 0.2, and "no correlation" when it is ± 0.2 or less.

Regarding the correlation coefficient, the condition that can calculate the regression equation is n-p-1 > 0 if the number of variables x is p and the number of samples is n. Only those satisfying this condition were calculated and shown in Table 9.3.

Table 9.3: Relationship between the hole location and the number of hole occurrence.

Type of accident	Organization		Local workplace	
	r	Regression equation	r	Regression equation
All marine accidents	−0.9	$y = -12.3x + 49$	−0.6	$y = -10.3x + 55.7$
Collision	−0.9	$y = -4.5x + 18.5$	−0.5	$y = -6.6x + 33.2$
Contact	−	−	0.8	$y = 1.4x - 2.4$
Grounding	−	−	0.3	$y = 0.4x + 2$
Occupational casualties	−0.9	$y = -6.2x + 23$	−0.8	$y = -4.1x + 17.1$

9.3.2 Correlation coefficient

In the organization, the correlation coefficient of all marine accidents, collisions, and occupational casualties was −0.9, and a very strong negative correlation was confirmed. It was recognized that the number of holes opened was larger in the first half of the process of the PDCA cycle and decreased in the latter process, and this trend was remarkable. This result shows that when the organization takes measures to prevent accidents, especially collisions and occupational casualties, the process in the early stages of the PDCA cycle, "plan" and "do", is important and that much effort needs to be directed towards these processes at organizations (Fig. 9.1). The correlation coefficient of all marine accidents in local workplaces was −0.6, and was strong correlation. It shows that, in general, operators in local workplaces need to take measures to prevent accidents during the early stages of the risk management process (Fig. 9.2).

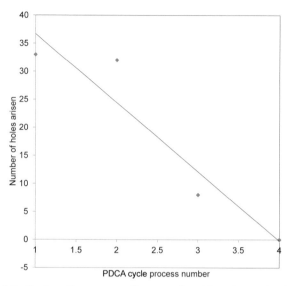

Fig. 9.1: Scatter diagram on all types of accident (organization).

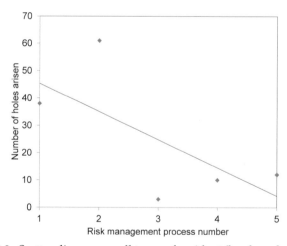

Fig. 9.2: Scatter diagram on all types of accident (local workplace).

Regarding each type of accident, in the local workplace, the correlation coefficient of collision was –0.6, and a strong negative correlation was confirmed. The same thing was said as in the organization. This result shows that when the operator takes measures to prevent collisions, it

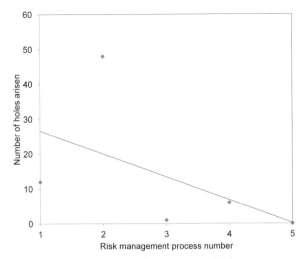

Fig. 9.3: Scatter diagram on collisions.

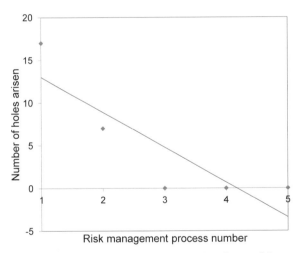

Fig. 9.4: Scatter diagram on occupational casualties.

is necessary to focus a great amount of effort on risk identification and risk analysis in the early stage of risk management process (Fig. 9.3). Concerning occupational casualties, a very strong negative correlation was confirmed (Fig. 9.4). Although the mode of occurrence of occupational casualties is greatly different from the collisions, both types of accident

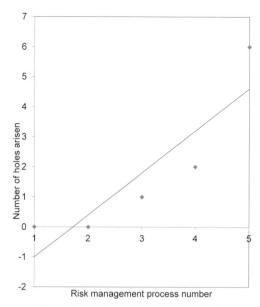

Fig. 9.5: Scatter diagram on contact.

have similarities; a great amount of effort should be focused on risk identification and risk analysis in order to prevent accidents.

Contact, unlike other types of accidents, had a very strong positive correlation, and the number of holes opened was small in the first half of the risk management process and increased in the latter period, and it was recognized that this tendency is prominent. This result indicates that when the operator takes measures to prevent a contact, much effort needs to be directed toward the later stage of risk management process, particularly monitoring and review (Fig. 9.5).

The correlation coefficient of grounding was +0.3, and it was confirmed that there was no correlation or weak positive correlation even if there was any correlation. The result indicates when the operator takes measures to prevent grounding, it is necessary to make efforts toward each stage of risk management process (Fig. 9.6).

9.4 Quantification accident model

When all the multiple regression equations and scatter diagrams in local workplaces and organizations are put into one integrated model, Fig. 9.7

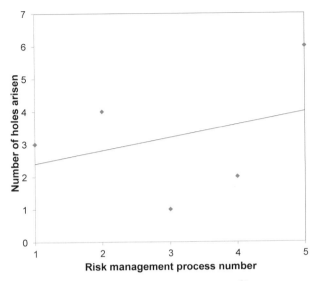

Fig. 9.6: Scatter diagram on grounding.

appears. The figures are produced based on analysis of 89 cases of marine accidents investigation reports issued by JTSB between 2008 and 2015, which do not include a marine incident.

It shows locations of the most frequent holes in local workplaces and organizations, latent conditions related to the accidents, tendency of locations of holes at local workplace and organizational level, respectively, and the numerical effect of latent conditions on accident occurrence described by the multiple regression equation.

Operators and organizations are able to see contributing factors which helped holes open and then lead their vessels to accidents. What is more, they can know how to prevent an accident. Tendency of location of holes described on scatter diagram is that, as a whole, holes tend to open during early stages of the risk management process at local workplaces and the same during the PDCA cycle at organizations. In more detail, when looking at the accident model, the most frequent holes tend to open during the risk analysis process in local workplaces and during the "Plan" process of the PDCA cycle in organizations. The multiple regression equation further shows that accidents occur due to such latent conditions in descending order: Software, peripheral liveware, hardware, environment and central liveware. Software and peripheral liveware are the most influential on occurrence of accident by order of magnitude, compared to hardware,

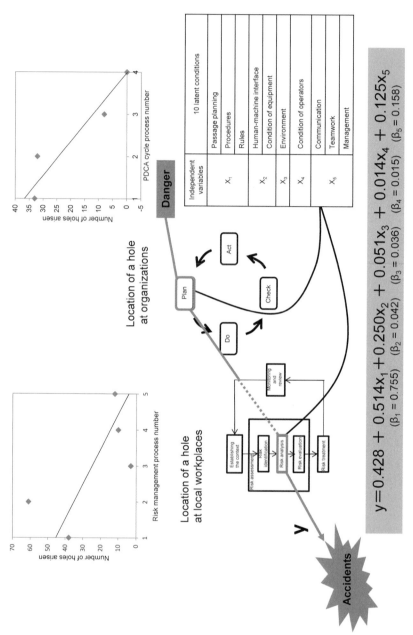

Fig. 9.7: Quantification accident model.

environment and central liveware. Software includes passage planning, procedures and rules; peripheral liveware involves communication, teamwork and management. Those latent conditions are areas to which organizations need to pay greater attention, make more efforts and ensure that there is no deviation from that which was originally planned or established at local workplaces and organizations.

9.5 Limitations of the study

There are two limitations to this study. First, some of the marine accident investigation reports did not include organizational factors. This research used the contents of the reports to classify the elements of the SHEL model; therefore, the findings, especially in relation to organizational factors, were not comprehensive, and the importance of peripheral liveware might have been underestimated.

Second, few cases involving fire, explosion, sinking or capsizing were analyzed in this research, and the sample size in these cases was small. This might have had an effect on the results in relation to the elements of the SHEL model and the accuracy of analysis.

9.6 Conclusions

From data of 89 cases of serious marine accident, the relation between latent conditions and number of marine accident occurrence became quantitatively clear, the points to pay attention to and efforts that operators and organizations need to make were indicated numerically. As shown in this chapter, the latent conditions can be identified in advance, and then there is a possibility that the multiple regression equation can be used to predict the accident occurrence, and people can effectively take measures for the prevention of accidents and against the recurrence of accidents.

Chapter 10

Convergence of Accident Models

10.1 Introduction

The accident models are divided into three types: Sequential, epidemiological and systemic. The latter two models are applicable to accidents occurring in modern socio-technical systems. The SCM, which represents the epidemiological accident models, was developed to explain every organizational accident.

This study clarified the characteristics of holes and the relationship between holes and latent conditions by using a risk management and process approach established in quality management systems. The study solved the problems relating to the SCM, and allowed for the development of the accident model using the Risk Management and Quality Management Process approach (RMQMP model) (Fukuoka 2016a, 2016b, Fukuoka and Furusho 2017).

The Systems-Theoretic Accident Model and Processes (STAMP) model, which represents the systemic accident model developed in missile safety systems, can be applied to loss of satellites, friendly fire accidents and bacterial contamination of a public water supply (Leveson 2011). The epidemiological accident model is applicable to accidents in the first classification quadrant; this category relates to marine transport, rail transport, power grids and dams. The systemic accident model is applicable to accidents in the second classification quadrant, which includes nuclear power plants, chemical plants, space missions, airlines and airports. The epidemiological accident model can only be applied to the first quadrant, whereas the systemic accident model can be applied to both the first and second quadrants (Perrow 1999, Hollnagel and Speziali 2008).

A significant difference between the models is the existence of a cause-effect link. The epidemiological accident model has a link, but the systemic accident model does not, and accidents occur when coordination among the components comprising the system changes over time (Leveson 2011, Hollnagel 2004).

This chapter aims to find the similarities and differences between the two types of accident models by applying them to an actual marine accident published by the MAIB. A sample report was selected that outlined the investigation into the capsizing and sinking of the cement carrier *CEMFJORD* in the Pentland Firth, Scotland, with the loss of all eight crew on 2 and 3 January 2015 (MAIB 2016). The choice was made at random from recent, very serious marine accidents in which local workplace and organizational factors needed to be included in the marine accident investigation reports; this was by provision of the Casualty Investigation Code (IMO 2008).

Although accident preventive measures or recommendations drawn from the analysis of each model are included in the analytical method, as shown below in Section 10.3, this study does not list them from the accident investigation report; this is because the purpose of this study is to clarify the similarities and differences in analytical methods between two types of accident models.

10.2 Summary of the sample accident

The 1,859 gross tons cement carrier *CEMFJORD* berthed at Aalborg Portland's cement loading terminal in Rordal, Denmark, and starting loading cement at 22:50 on 29 December 2014. Loading was occasionally suspended due to the malfunction of the port ballast pump, which caused the vessel to list to port, but resumed by installing a portable submersible pump to aid de-ballasting.

At 13:00 on 30 December, fully loaded with 2,084 tons of cement, the *CEMFJORD* sailed for Runcorn on the west coast of the UK. The vessel's average speed over the ground (SOG) from Rordal to the Skagerrak, the entrance of the North Sea, was 9.2 knots. After passing the Skagerrak, the *CEMFJORD* proceeded across the North Sea toward the Pentland Firth, a narrow strait between the Orkney Islands and the north of Scotland, at average SOG of about 6 knots. The UK Meteorological Office issued deteriorating weather forecasts in the areas that covered Pentland Firth, stating west wind 7 to severe gale wind 9 (wind speed 41–47 knots), very rough or high in north.

At 10:52 on 2 January 2015, in the eastern approaches to the Pentland Firth, the master reported the vessel's position, intended port of call and details of cargo to the Shetland Coastguard over VHF radio in accordance with the Pentland Firth voluntary reporting scheme. The coastguard monitored the progress of the *CEMFJORD* on AIS but there was no requirement for the master to submit an exit report.

At 13:15, the *CEMFJORD*'s AIS transmissions ceased. At 14:16 on 3 January, the capsized *CEMFJORD* was discovered by a roll-on roll-off passenger ferry. Search and rescue operations commenced, but all eight crew were missing.

The accident investigation report concluded that the *CEMFJORD* capsized in extraordinarily adverse sea and weather conditions; it also concludes that the conditions were predictable and could have been avoided by passage planning, and that the rapid nature of the capsize prevented the crew from sending a distress message or abandoning the vessel. It also determined that the *CEMFJORD* proceeded to sea with significant safety regulation deficiencies, as a result of safety regulation exemptions approved by the flag state. The accident went unnoticed for 25 hours, until a passing ferry found the upturned hull of the *CEMFJORD*; the delay occurred because the Shetland Coastguard of the Maritime and Coastguard Agency did not require an exit report when the vessel left the voluntary reporting scheme area in the Pentland Firth.

10.3 Analytical method

The accident investigation report into the capsizing and sinking of the cement carrier *CEMFJORD* was analyzed using processes in both the RMQMP and STAMP models.

The analysis process in the RMQMP model is as follows:

(1) The definition of a hole is determined. This means that an unacceptable risk, as stated by ISO/IEC Guide 51 (ISO/IEC 1999), exists in an organization or a local workplace. In this case, the vessel capsized; therefore, a hole opened when the vessel's stability was reduced to the extent that led to the capsizing.

(2) Locations of holes in a local workplace and in an organization are identified, using the risk management process and the PDCA cycle, respectively. To determine the locations of holes in a local workplace, the procedures for passage planning defined by Swift (2000) and IMO (2000b) are used.

(3) Latent conditions that caused the opening of holes and led to the accident are classified into 10 groups, in accordance with their definitions. The ten latent conditions are: Passage planning, procedures, rules, human-machine interface, condition of equipment, environment, condition of operators, communication, teamwork and management. Analyses of the locations of holes and latent conditions are conducted with all organizations related to the accident; for each organization, latent conditions are written in a textual form in a table.

(4) The locations of holes and related latent conditions are shown graphically.

(5) Accident preventative measures are used to shut the holes. Each latent condition is rectified by implementing the methods of risk reduction stated in ISO/IEC Guide 51, and risk reduction is prioritized as follows: Inherently safe design, protective devices, information for safety, additional protective devices, training, personal protective equipment and organization.

For the STAMP model, the analysis process is as follows:

(1) The systems and hazards involved in the loss are identified.

(2) The system safety constraints and system requirements relating to the hazards in the safety control structure are determined.

(3) The loss at a physical system level is analyzed. Factors contributing to ineffective physical and operational controls, physical failures, dysfunctional interactions, communication and coordination flaws, and unhandled disturbances are all considered.

(4) At higher levels of the safety control structure, the inadequate control that led to the contributing factors is determined. Assignment of responsibilities and inadequate enforcement of assigned responsibilities, context and influences on the decision-making process, and flaws in the mental models of those making the decisions are all considered.

(5) Coordination and communication related to the loss are examined, and dynamics and changes in the system and the safety control structure over time are determined.

(6) Safety requirements and constraints, the context in which decisions are made, inadequate control actions, and mental model flaws are each described in textual form in a table of each component. Control channels and communication channels are described by arrows between the components that constitute the system.

(7) Recommendations are generated. There is no algorithm for identifying the relative importance of recommendations (Leveson 2011).

In this case, the accident investigation report determined that, even if a distress alert from an Emergency Position Indicating Radio Beacon (EPIRB) or a distress radio call from the *CEMFJORD* had been raised at the time of the capsize, the rapid nature of the capsize and ferocious sea conditions meant the outcome for the vessel and crew would almost certainly have been the same. Therefore, in this study, the analysis focused on factors that led to the capsize, and in both models, other factors were written in parentheses.

10.4 Limitations of analysis

When the accident investigation report did not contain any organizational factors or local workplace factors, the study could not analyses holes or latent conditions in the RMQMP model and safety requirements and constraints, the context in which decisions were made, inadequate control actions, or mental model flaws in the STAMP model.

10.5 Results

10.5.1 Analysis using the RMQMP model

Figure 10.1 illustrates the latent conditions derived from Table 10.1, and the locations of holes that opened at the local workplace, the *CEMFJORD*, and the organization (Brise Bereederungs GmbH). The master decided to proceed to the Pentland Firth despite very poor conditions arising from a maximum westerly current being opposed by westerly gale force winds. Therefore, a hole opened during the execution of the passage planning or the risk treatment during the risk management process.

Brise Bereederungs GmbH did not take corrective action regarding passage planning, loading operations, or the vessel's stability after the

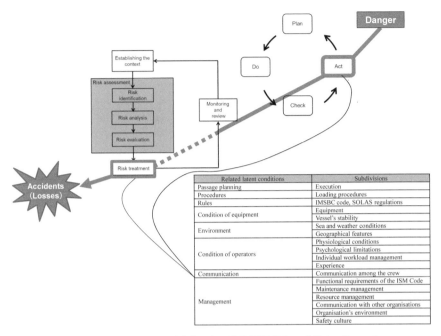

Related latent conditions		Subdivisions
Passage planning		Execution
Procedures		Loading procedures
Rules		IMSBC code, SOLAS regulations
Condition of equipment		Equipment
		Vessel's stability
Environment		Sea and weather conditions
		Geographical features
Condition of operators		Physiological conditions
		Psychological limitations
		Individual workload management
		Experience
Communication		Communication among the crew
Management		Functional requirements of the ISM Code
		Maintenance management
		Resource management
		Communication with other organisations
		Organisation's environment
		Safety culture

Fig. 10.1: The RMQMP model for the *CEMFJORD* and Brise Bereederungs GmbH, showing holes and latent conditions.

investigation into a separate cargo shift incident occurred on 7 October 2014; they also failed to rectify the cargo hold bilge-pumping system. Instead, they repeatedly sought safety regulation exemptions from the flag state. A hole therefore opened during the 'act' process of the PDCA cycle.

The hole that opened at the ship management company, Brise Bereederungs GmbH, was influenced by the way that the vessel's flag state, the Department of Merchant Shipping for the Republic of Cyprus (DMS Cyprus), approved the exemptions. Approvals were granted without understanding the nature of the work undertaken and the new level of risk caused by the work on board the *CEMFJORD*. Because the accident investigation report did not describe in detail who at DMS Cyprus was responsible for conducting exemption risk assessments, the location of a hole on the part of an operator was not analyzed; thus, the RMQMP model for DMS Cyprus, which could connect with the 'act' process of the RMQMP model for the *CEMFJORD* and Brise Bereederungs GmbH, is not shown in Fig. 10.1.

Table 10.1: Analysis using the RMQMP model for the *CEMFJORD*, and Brise Bereederungs GmbH.

10 latent conditions	Subdivisions	Details of latent conditions which led to the accident
Passage planning	Stages of appraisal, planning, execution, and monitoring	Regarding the stage of execution, the Cemfjord approached the Pentland Firth at the worst possible time, with the maximum westerly current being opposed by westerly gale force winds. In this case, an alternative plan such as seeking shelter, slowing down, or diverting via the English Channel should be executed.
Procedures	Procedures, manuals, checklists, station bills, standing orders, company rules, and others	The cargo was settled unevenly, deviating from the loading procedures because of list to port by about 5 degrees, which subsequently increased the risk of the cargo shifting in heavy sea.
Rules	The convention on the international regulations for preventing collisions at sea, 1972, International convention on standards of training, certification and watchkeeping for seafarers, International convention for the safety of life at sea (SOLAS), local navigation rules, and others	The density of the cargo was not considered properly, resulting in parameters outside the international maritime solid bulk cargoes code (IMSBC code). Safety related shortcomings such as modification of life saving appliances (LSA) and defective bilge pumping system were exempted by the flag state. (Abandon ship procedures were not practiced.)
Human-machine interface	Design of work stations, displays, controls and other factors that constitute a human-machine interface	
Condition of equipment	Structure/machinery/equipment	Cargo hold bilge pumping system was defective. (A rescue boat could not be launched due to long lifting slings.) (Emergency position indicating radio beacon horizontally mounted on the bridge wing was not a float-free arrangement, resulting in it being trapped in the upturned hull.)
	Vessel's stability	The density value of the white cement was 1100 kg/m^3, and the vertical center of gravity was higher than the one in the loading manual; as a result, the IMO's minimum stability criteria were not satisfied.

Local workplace

Environment	Sea and weather conditions	The Cemfjord approached the Pentland Firth with the maximum westerly current being opposed by westerly gale force winds, resulting in a shift in the cement cargo as she heeled beyond 30 degrees.
	Conditions people are working in	
	Traffic density	
	Geographical features	The Pentland Firth is a channel where mariners can encounter extensive and dangerous conditions.
	Berth facilities and other factors	
Condition of operators	Physical or sensory limitations	
	Physiological conditions	Master and crew suffered fatigue or tiredness resulting from by cargo loading problems and deteriorating sea conditions.
	Psychological limitations	Master was under pressure due to the delayed departure from Rordal and further delays due to the weather in the North Sea.
	Individual workload management	Tight charterer's schedule prevented master and crew from taking time for routine maintenance.
	Knowledge, skill, experience, education/training	Recent master's experience of a near miss when the cargo shifted during a turn in heavy sea, led to his unwillingness to alter course.
Communication	Communication among the bridge team, between a pilot and the bridge team, or between the bridge and vessel traffic services	There was no challenge against the master's operational decisions because of lack of experience of chief officer and crew.
Teamwork	Roles and responsibilities of the crew, pilot, and other people involved in an accident	

Table 10.1 contd. ...

...Table 10.1 contd.

10 latent conditions	Subdivisions	Details of latent conditions which led to the accident
Management	Functional requirements of ISM Code regarding safe operation	Corrective actions regarding passage planning, loading operations and vessel's stability were not taken after the investigation into the cargo shift incident on 7th October 2014.
	Maintenance management	Rectifications of the LSA and cargo hold bilge pumping system were not taken; instead, safety regulation exemptions were repeatedly sought to the flag state.
	Emergency preparedness	(Specific abandon ship procedures were not established.)
	Resource management	A stability computer was absent. Lack of resource management was dominant because of the policy of running equipment until it failed.
	Communication with ship or within the organization	
	Communication with other organizations (Flag states, recognized organizations, manning companies, etc.)	Misleading messages were given to the flag state and the recognized organization regarding safety regulation exemptions by phone call or email.
	Organization's environment	Commercial pressure existed, leading top management to seek safety regulation exemptions repeatedly.
	Safety culture (informed culture)	Lessons learned were not used.

Organization

Table 10.2: Analysis using the RMQMP model for the DMS Cyprus.

10 latent conditions	Subdivisions	Details of latent conditions which related to the accident
Procedures	Local procedures at Shetland Coastguard	Exit reports from vessels were not required for the purpose of reducing levels of very high frequency (VHF) radio traffic. No alert system was established when the automatic identification system (AIS) transmission from a vessel ceased.
Rules	SOLAS regulations, IMO resolutions	The purpose of the voluntary reporting scheme was not defined. Distinctions between mandatory or voluntary reporting schemes were not offered.
Condition of equipment	Maintenance of equipment	The operation room data distribution system was faulty, and the AIS information was not displayed on the screen.
Teamwork	Roles and responsibilities of the crew, pilot, and other people involved in an accident	Watch officer was assigned the task of monitoring VHF radio traffic and responding to vessel's maritime reports, not monitoring vessel's progress.
Management	Requirements of safety management system based on the quality management system	It was not prepared for emergency situations, because no measures of identifying AIS transmission failures existed.
	Organization's environment	IMO did not offer any distinction between 'mandatory' or 'voluntary' regarding ship reporting systems.

Table 10.1 shows the latent conditions that led to the accident regarding the *CEMFJORD* and Brise Bereederungs GmbH. Tables 10.2 and 10.3 show latent conditions relating to the DMS Cyprus, and the Shetland Coastguard of the Maritime and Coastguard Agency. Holes and latent conditions for the organizations, Det Norske Veritas-Germanischer Lloyd, and Lloyd's Register were not analyzed because these were not described in the accident investigation report.

10.5.2 Analysis using the STAMP model

For the *CEMFJORD* accident, Table 10.4 shows each component that constitutes the system. The components were IMO, the DMS Cyprus,

Table 10.3: Analysis using the RMQMP model for the Shetland Coastguard.

10 latent conditions	Subdivisions	Details of latent conditions which related to the accident
Rules	SOLAS regulations	The CEMFJORD, with safety regulation exemptions, was allowed to proceed to sea with safety deficiencies relating to rescue boat launching arrangements and cargo hold bilge pumping system.
Management	Requirements of safety management system based on the quality management system	An effective mechanism to identify different opinions in the inspections between its surveyors and PSC officers was not established. Managing the exemptions was not established.
	Communication with Brise Bereederungs GmbH and Det Norske Veritas-Germanischer Lloyd	All communication was done by phone call or email, and the information passed was ambiguous and misleading concerning safety regulation exemptions.
	Organization's environment	Global industry pressure existed, leading management to approve the safety regulation exemptions without a real understanding of the situation onboard.

the Shetland Coastguard of the Maritime and Coastguard Agency, Det Norske Veritas-Germanischer Lloyd, Lloyd's Register, Brise Bereederungs GmbH, and the *CEMFJORD*. For each component, it describes the safety requirements and constraints violated, the context in which decisions were made, inadequate decisions and inadequate control actions, and mental model flaws in each component; some components are interrelated. The IMO establishes safety regulations, which are enforced by the flag states; the flag states ensure that the flagged vessels and ship management companies are following the regulations both by themselves and through recognized organizations.

Table 10.4 also illustrates that misleading requests for safety regulation exemptions on the *CEMFJORD*, sought from Brise Bereederungs GmbH, were sent to the DMS Cyprus; those requests were not scrutinized within the DMS Cyprus and Det Norske Veritas-Germanischer Lloyd organizations. Furthermore, communication between the DMS Cyprus and the *CEMFJORD* did not take place, and safety regulation exemptions were approved by the flag state without clarifying their ramifications. It also illustrates that the Shetland Coastguard did not require an exit report; therefore, the communication channel was unilateral, implying

Table 10.4: Analysis using the STAMP model.

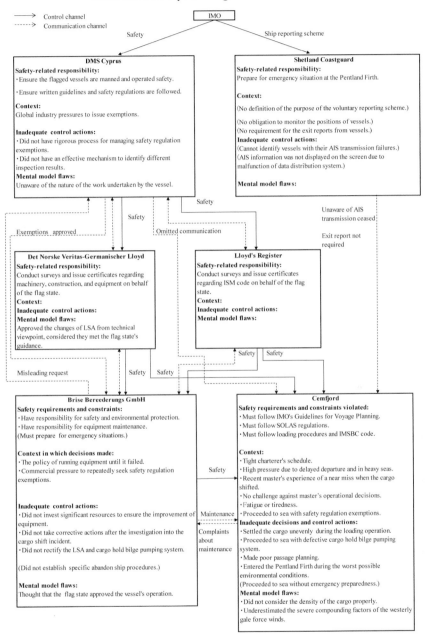

that it could not detect that the AIS transmission from the *CEMFJORD* had ceased while passing through the Pentland Firth.

10.5.3 *Differences and similarities among the two models*

The contents of Tables 10.1 to 10.3 are almost equivalent to those in Table 10.4, although the representation of the analysis is different. This means that results of the accident analyses by the two different types of accident models are the same. This finding is further supported when the analytical methods of the RMQMP and STAMP models are compared. Comparison of the analytical methods is as follows:

(1) A hole on the RMQMP model is equal to the hazard involved in the loss on the STAMP model.

(2) Local workplaces and organizations are identical in the components that constitute the system.

(3) Risk management embedded at local workplaces, as well as the PDCA cycle embedded in organizations comprise the system safety constraints or controls in the STAMP model.

(4) The safety management system or quality management system is the safety control structure in the STAMP model.

(5) Ten latent conditions are equivalent to contributing factors to ineffective physical and operational controls, physical failures, dysfunctional interactions, communication and coordination flaws, unhandled disturbances, and so forth on the STAMP. The reason for this equivalency is that 10 latent conditions are mostly drawn from the concept of the SHEL model, which explains the interrelationship among five factors: Software, hardware, environment, central liveware and peripheral liveware.

(6) The latent conditions of passage planning, procedures and rules in local workplaces and deviation from the safety management system in the RMQMP model mean dynamic change in the system and in the safety control structures in the STAMP model over time.

Regarding representation of the system and the failure of system control, each model takes on a different style.

(1) The RMQMP model can depict the whole system both in textual and graphic form by adding the risk management and safety management system defenses of the organizations involved in the accident. In

contrast, the STAMP model can depict the whole system, with control channels and communication channels in a single table.

(2) The RMQMP model indicates the locations of unacceptable risk in the risk management and safety management system defenses in graphic form. In the STAMP model, the location of risk is described within the inadequate decision and control action description in textual form.

10.6 Convergence of different type of model

Underwoods and Waterson (2014) stress the importance of the system-thinking approach in accident analysis in modern socio-technical systems. This approach includes the system structure, system component relationship and system behavior, which can be described as follows: The hierarchical level of the system and its boundaries are indicated in the system structure; all components and their interrelationship are considered in the system component relationship; the way that goals, resources and environmental conditions influence human behaviors are determined in the system behavior. According to the comparison discussed above, both accident models include the system-thinking approach.

Accident models are divided into three categories, and each accident model has applicable fields based on the two dimensions of coupling and tractability (Hollnagel 2004, Hollnagel and Speziali 2008). The study found that the RMQMP model, which solved the disputable SCM questions, has the same analytical methods as the STAMP model, which is classified as a systemic accident model. This finding indicates the convergence of the epidemiological and systemic accident models and paves the way for the practical application of the RMQMP model to fields where only the systemic accident model is currently applicable.

10.7 Conclusions

Underwoods and Waterson concluded, after comparison of the epidemiological and systemic accident models, that the systemic accident model is not used by investigators or practitioners; the reason is that investigators have limited time for publishing reports, and it requires considerable time to investigate all aspects of past and current situations, for each component and contributing factor.

This chapter describes the convergence of the epidemiological and systemic accident models. The RMQMP model includes organizational factors based on ISO 9001 and 9004, and can graphically present weaknesses of a system. It can assist investigators and practitioners with limited time by providing the areas in which they should investigate, analyze, and establish preventive measures. When the definition of latent conditions is modified to fit other fields, the RMQMP can also help people working in these fields enhance safety.

Chapter 11

Rectification of the Weakness and Improvement of the System

11.1 Introduction

Rectification of weakness and improvement of the system is accomplished by systematic approach on accident prevention using the accident model, statistical science, risk management and the PDCA cycle. In this chapter, the process of systematic accident prevention, which incorporates those tools, is introduced in order to reduce the number of marine accidents and incidents, resulting in safer seas with fewer accidents (Fukuoka 2016b). The systematic approach can be applied to accidents and incidents; the later part of this chapter explains the difference between an accident and an incident occurrence.

11.2 Principles of the systematic accident prevention

The systematic accident prevention is needed for the following three reasons: (1) Accidents occur every day somewhere in the world, (2) Safety Management System has not demonstrated the effectiveness to reduce the number of marine accidents since its implementation in 1998, and (3) There is no integrated system to carry out accident prevention.

Wiegmann and Shappell (2003) state that it is necessary to clarify the hole in order for the SCM to be systematically used for accident analysis and accident prevention. Since unresolved issues on the SCM were clarified, as described in Chapter 8, the systematic accident prevention can be taken.

To implement the systematic accident prevention, some preconditions exist. Firstly, SMS is established and conducted at the organization. Secondly, risk management process is conducted at a local workplaces or vessel. Thirdly, people are familiar with the PDCA cycle and risk management process. Finally, when any accident or incident occurs the scientific accident investigation is conducted.

Those are the principles of the systematic accident prevention.

(1) Conduct scientific accident investigation.

An organization (ship management company) is supposed to conduct an accident investigation according to the ISM Code. The accident investigation needs to be based on evidence: AIS, VDR, physical evidence, documentary evidence and human evidence by proper interviewing techniques.

(2) Employ the accident model that fits to the shipping industry.

The epidemiological accident model is applicable to accidents in the marine domain.

(3) Visualize a weakness in the system.

Visualization of a weakness (holes and latent conditions) in the system is performed by using the RMQMP model. Once visualized, it is easy to control the weakness.

(4) Use the risk management process for risk reduction.

The weakness is rectified by using the risk reduction method according to the hierarch of controls.

(5) Utilize statistical science on accidents and incidents.

In order to customize the accident prevention to fit the organization, the frequent accidents, holes and latent conditions are identified by utilizing statistics accumulated by the organization.

(6) Put emphasis on concept of risk management and the PDCA cycle (the core of the SMS), optimize the resources to be allocated to reduce the number of accidents and incidents.

At first, the number of frequent accidents is reduced. As a result, non-frequent accidents remain. After that, they are prevented by using the same systematic method.

11.3 Process of systematic accident prevention

Figure 11.1 illustrates the process of systematic accident prevention. It consists of recurrence prevention after an accident or an incident occurs and proactive measures which can be taken by studying and analyzing accident reports produced in the past by own company (Fukuoka 2016b).

11.3.1 Recurrence prevention

To prevent accidents from recurring, it is necessary to take measures in the following procedures.

(1) When investigating accidents and incidents that occurred in own company, exhaustively collect evidence related to the accidents using the SHEL model.

(2) Use ECFC to find event chain and contributing factors.

(3) Use risk management and the PDCA cycle to clarify the location of hole opened and identify the latent conditions related to the accident and incident according to the classification of latent conditions shown in Chapter 8.

(4) Draw RMQMP model when holes and latent conditions are identified.

(5) Rectify all holes and latent conditions related to the accident and incident using risk reduction methods shown in Chapter 2 to prevent recurrence.

(6) Incorporate rectified holes and latent conditions into previous SMS and revise it.

(7) Conduct education and training in accordance with the revised "Plan" in SMS.

(8) When resuming ship operation and cargo operation, utilize RMQMP model applied to own company and abstract generalizations (general characteristics of accidents) in Chapter 8, and be aware of the impact of latent conditions on accident and incident occurrence discovered in multiple regression equation in Chapter 9.

(9) Regularly confirm that measures shown above (6) are being implemented at organization and local workplace, and executes the PDCA cycle to ensure safety.

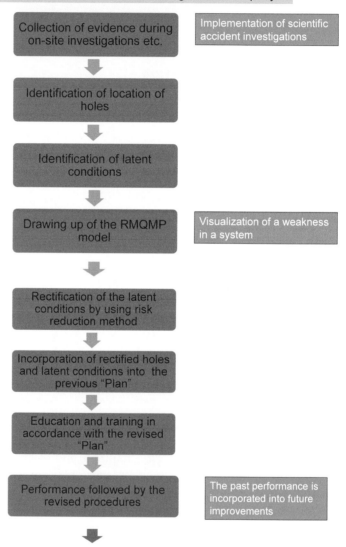

Recurrence prevention

Occurrence of accidents and incidents relating to own company

Collection of evidence during on-site investigations etc.

Implementation of scientific accident investigations

Identification of location of holes

Identification of latent conditions

Drawing up of the RMQMP model

Visualization of a weakness in a system

Rectification of the latent conditions by using risk reduction method

Incorporation of rectified holes and latent conditions into the previous "Plan"

Education and training in accordance with the revised "Plan"

Performance followed by the revised procedures

The past performance is incorporated into future improvements

Resumption of ship operation and cargo operation

Fig. 11.1 contd. ...

...Fig. 11.1 contd.

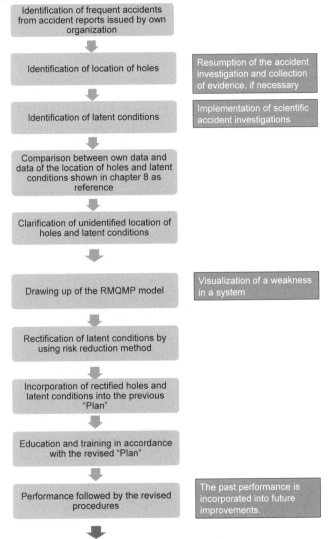

Continuation of ship operation and cargo operation

Fig. 11.1 contd. ...

...Fig. 11.1 contd.

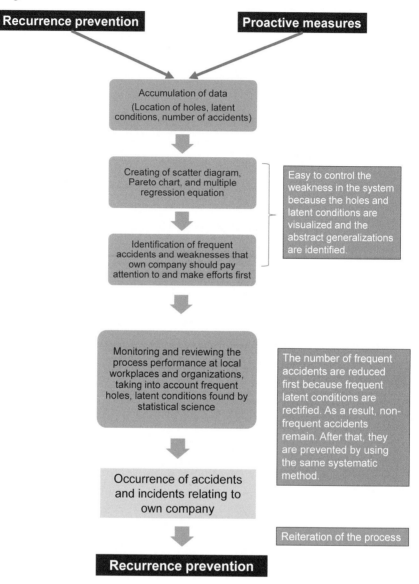

Fig. 11.1: Process of systematic accident prevention.

11.3.2 *Proactive measures*

Heinrich (1959) points out that safety management is similar to product quality control. According to Iizuka (2009), quality control utilizes the Pareto chart when choosing problems to be managed or defect items of products. As it is assumed that the resources available to the organization are limited, it is necessary that proactive measures are taken in the following procedures, including utilization of the Pareto chart.

(1) Identify accident and incident types that are occurring frequently from statistical data in own company.

(2) For each accident and incident, find the location of holes and latent conditions using accident and incident reports investigated by own company.

(3) Compare the location of holes and the latent conditions found in (2) with the location of holes in the organizations and local workplaces and latent conditions described in Chapter 8 by the types of accident. Then, clarify the unidentified holes and latent conditions. If holes and latent conditions that were not specified in own company's accident report exist, the location of the hole and latent conditions are reviewed and added.

(4) Draw RMQMP model when holes and latent conditions are identified.

(5) Rectify all latent conditions related to the accident using risk reduction methods.

(6) Incorporate rectified holes and latent conditions into previous SMS and revise it.

(7) Conduct education and training in accordance with the revised "Plan" of the PDCA cycle in SMS.

(8) Monitor the performance comparing with revised "Plan" and "Do" in SMS.

(9) Accumulate the data on the location of holes, latent conditions, and the number of accidents and incidents related to own company, including the data investigated in recurrence prevention part.

At the stage when own company's data of accidents and incidents has accumulated almost 100 cases or more, it is necessary to use statistical science to create a scatter diagram, Pareto chart, and the multiple regression equation that can show the direction in which own company's efforts should be paid to enhance safety effectively and efficiently.

(10) Create scatter diagram, Pareto chart, and the multiple regression equation from the latent conditions and data on the number of accidents obtained in (9). Then, produce a quantification accident model which consists of the RMQMP model, scatter diagram and the multiple regression equation applied to own company. As a result frequent holes and latent conditions in your own company are revealed.

(11) Regularly confirm that measures taken regarding these holes and latent conditions shown above (10) are being implemented in the organization and the local workplace, executing especially the "Check" of the PDCA cycle in SMS in order to ensure safety. At this stage, the number of frequent accident and incident occurrence will decline because all measures to prevent them have already been taken. The whole process of systematic accident prevention is based on the concept of "vital few, trivial many", which means that, as the process goes through, trivial events to start with become vital events, and then even the number of non-frequent incidents will decline.

(12) When if an accident or an incident occurs, reiterate the process of recurrence prevention shown in 11.3.1.

11.4 An accident and an incident

Until Chapter 10 in this book, the main subjects to be dealt with were matters related to accidents. Accidents and incidents are differentiated in Casualty Investigation Code 2.9 and 2.10 as follows.

Marine accident or casualty means an event, or a sequence of events, that has resulted in any of the following which has occurred directly in connection with the operations of a ship: (1) the death of, or serious injury, to a person; (2) the loss of a person from a ship; (3) the loss, presumed loss or abandonment of a ship; (4) material damage to a shop; (5) the stranding or disabling of a ship, or the involvement of a ship in a collision; (6) material damage to marine infrastructure external to a ship, that could seriously endanger the safety of the ship, another ship or an individual; (7) severe damage to the environment, or the potential for severe damage to the environment, brought about by the damage of a ship or ships.

A marine incident means an event, or sequence of events, other than a marine casualty, which has occurred directly in connection with

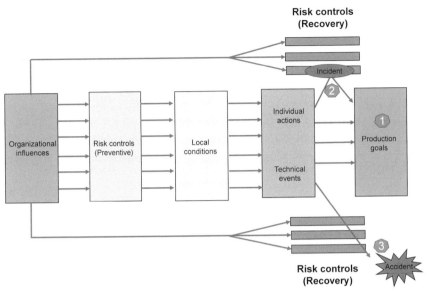

Fig. 11.2: Accident and incident occurrence mechanism (Adapted from ATSB 2008).

the operations of a ship, that endangered, or, if not corrected, would endanger the safety of the ship, its occupants or any other person or the environment.

The incident does not develop into an accident due to the function of defensive layer, and its occurrence mechanism is different from the one of accident. Figure 11.2 shows how an accident and an incident occur using one of ATSB models derived from Middle SCM in Chapter 3. An organization achieves its production goals through the combination of various events and conditions, as illustrated in 1 of the Fig. 11.2. In some cases, events and conditions will combine to produce accident events, depicted by the angled arrows. An incident occurs when a risk control works for minimizing the consequence, as shown in 2, whereas an accident occurs when a risk control does not work, as shown in 3.

In this section, the study clarifies how the difference of occurrence mechanism between an accident and an incident will work and appear as events and contributing factors. First, a case of incident between the *AUTO BANNER* and the *SHIMAYUKI* (JTSB 2014e) is introduced, and then collision between the *YONG SHENG VII* and the *HOKUEI No. 18* (JTSB 2016a).

Incident between the *AUTO BANNER* and the *SHIMAYUKI*

Summary: The *AUTO BANNER*, Panama flagged 52,422 gross tons car carrier, was proceeding south with a pilot onboard along in the right-side of the Kanmon Passage, which lay on the starboard side of the *AUTO BANNER*, while the *SHIMAYUKI*, Japan flagged 3,050 gross tons training vessel, was proceeding north on the left-side of the same passage which lay on the port side of the *SHIMAYUKI* while altering her course, and then both vessels approached to close quarter situation. The *AUTO BANNER* turned to port to avoid collision, passing starboard to starboard with the *SHIMAYUKI* by 250 m, at 2:050 on 11 June 2013.

Events and contributing factors

Maneuvering of the *AUTO BANNER*

The *AUTO BANNER* had departed from the Port of Hakata in Japan, bound for the Port of Hanshin through the Kanmon channel, and a pilot for the Kanmon Passage embarked at about 1.2 miles northern part of Mutsure shima at 20:40 on 11 June. The pilot looked through the pilot card and made notes on necessary items, such as particulars and maneuverability of the *AUTO BANNER*. He observed a vessel nearby sailing south (hereafter referred to as "vessel C") and another vessel, the *SHIMAYUKI*, sailing north in the Kanmon Passage with radar and AIS, and directed to proceed south into the entrance of the Kanmon Passage at a speed of 13 knots.

The pilot noticed the *SHIMAYUKI* sailing along near the middle of the the Kanmon Passage at 20:43 to 20:44. Meanwhile, the *AUTO BANNER* was altering her course to starboard along the passage, the pilot felt doubt about the movement of the *SHIMAYUKI* when he saw the *SHIMAYUKI* altering her course to port by judging from the change in appearance of navigation lights.

At 20:48, the pilot perceived sidelights of the *SHIMAYUKI* at starboard 20°, 1,000 m forward, thinking that if the *AUTO BANNER* had continued turning to starboard, both vessels would face the danger of collision. After checking the condition of other meeting vessels and the position of the No. 5 light buoy, the pilot instructed to take hard port, and sound a maneuvering signal of two short blasts towards the *AUTO BANNER*. The pilot did not use VHF communication with the *AUTO BANNER* as there was not enough time left.

The pilot instructed to sound two short blasts repeatedly until the *AUTO BANNER* started turning to port. When the *AUTO BANNER* turned to port, he ordered the master of the *AUTO BANNER* to emergency stop engine and take the helm to midship.

The pilot perceived the shadow of the *SHIMAYUKI* when both vessels approached about 700 to 800 m. At 20:49, the pilot observed the situation that the *SHIMAYUKI* was turning to starboard, and he strongly felt a danger of collision. After acknowledging the *SHIMAYUKI* turning to port again, the pilot heard two short blasts being sounded by the *SHIMAYUKI*. At 20:50, both vessels passed by about 250 m away.

Maneuvering of the *SHIMAYUKI*

The chief navigator drafted the passage planning on the Kanmon channel which was intended to pass 250 m inward from the East boundary line of the Kanmon Passage. The master approved the passage planning prior to sailing, and the *SHIMAYUKI* departed from the Port of Kure in Japan, bound for the Port of Sasebo through the Kanmon channel, at 10:10 on 11 June.

At 20:40, lookouts and a radar watchman recognized the presence of the *AUTO BANNER* and a vessel C which was sailing in the same direction, derived from information of AIS and radar, and reported it to the master and the chief navigator, but the chief navigator did not hear the information about the *AUTO BANNER*.

At 20:43, while their own vessel was heading at 033°, the master and the chief navigator received a report from a crew "(*SHIMAYUKI* is proceeding at), 30 m left from the planned track." The chief navigator thought that their own vessel had been sailing as planned, judging by clearing bearings related to a headmark. The chief navigator did not realize that their own vessel was sailing near the middle of the Kanmon Passage while being affected by wind and tidal streams. As their own vessel was approaching the next way point, the chief navigator decided not to modify the course but to alter the course at the way point, which was approved by the master.

At 20:47 the chief navigator heard a report of "course alteration" from the crew and assumed that their own vessel was either about to arrive at the way point or had already reached it. In addition, the chief navigator felt very close to the line connecting the No. 5 light buoy and the No. 3 light buoy on the East boundary line of the Kanmon Passage, which was near the way point. From this situation, the chief navigator instructed the helmsman to steer to port 15° and the vessel started turning to the left. When their own vessel was turning to 027°, the master saw the vessel C's sidelights changing from green light to red light, recognized the distance from the vessel C was not sufficient, and then instructed the chief navigator to take starboard 15°. At this time the master and the chief navigator first realized the presence of the *AUTO BANNER* sailing behind the vessel C.

The master heard a maneuvering signal of two short blasts but did not recognize that it was sounded by the *AUTO BANNER*. The master heard the second maneuvering signal of the two short blasts at about 732 m away from the *AUTO BANNER* at 20:48 and received the same report from a crewmember, then he acknowledged that the signals came from the *AUTO BANNER*.

The master recognized that the *AUTO BANNER* was turning to the left, intending to pass starboard to starboard. He instructed the chief navigator to take port 30° and sound two short blasts.

The master, while the *SHIMAYUKI* was turning to the left at the speed of 7.5 knots, heard another two short blasts from the *AUTO BANNER*, and then instructed to sound two short blasts. He found that CPA was about 247 m with radar when passing starboard to starboard with the *AUTO BANNER* at the speed of about 13 knots at 20:50.

These are contributing factors related to the incident on the *SHIMAYUKI*.

Passage planning

(1) The chief navigator had expected a westward tidal stream, which would weaken the steering effect while navigating the Kanmon Passage, and intended to have enough safe margin in case of taking collision avoidance action. In addition, the South breakwater light was selected as a headmark. As a result, the intended track was set about 250 m inward from the eastern boundary line of the Kanmon Passage.

(2) At 20:47, course alteration was made 500 m before the way point at the Kanmon Passage. Taking into account the ship's position that had already been about 30 m inward from the intended track around that time, the *SHIMAYUKI* was navigating across the middle passage, resulting in proceeding to the left side of the passage which lay on her port side. The chief navigator did not monitor the ship's position before and after the course alteration.

(3) From these facts, the process of monitoring and reviewing the passage planning was insufficient.

Rules

The Rules for Enforcement of the Port Regulation Law stipulate that steamships that navigate the Kanmon Passage must keep as near to the right side of the passage as possible.

Environment

(1) According to sailing directions published by the Japan Coast Guard, most waters of the Kanmon channel are designated as the port area of the Kanmon Port, within which the Kanmon Passage is established by the Port Regulation Law. There are many bends in the channel, the width of navigable waters ranges from 1 mile to 500 m. Tidal stream is strong, vessel traffic is extremely high.

(2) At the time of the incident, there was 3 knots westerly tidal stream.

(3) The passage width in the water near this incident occurrence place was about 750 m.

(4) Vessel C was sailing ahead of the *AUTO BANNER*, which contributed to the delay in the master and the chief perceiving the *AUTO BANNER*.

Condition of operators

(1) The master and the chief navigator had paid too much attention to vessel C prior to the incident; therefore, they did not recognize the movement of the *AUTO BANNER* until their own ship's course alteration at 20:47. Their situation awareness was incomplete.

Communication

(1) The master and the chief navigator had been informed by the crewmembers about the positions of own vessel and other vessels, including the *AUTO BANNER*. The chief navigator did not recognize own ship's position was about 30 m inward from the planned track.

(2) At 20:47, Kanmon Marine Traffic Information Service (MARTIS) issued the warning by VHF towards the *SHIMAYUKI* that a large vessel was approaching and should navigate the right side of the passage, but the crew of the *SHIMAYUKI* did not notice this warning.

(3) From these facts, communication among the bridge team and the Kanmon MARTIS was insufficient.

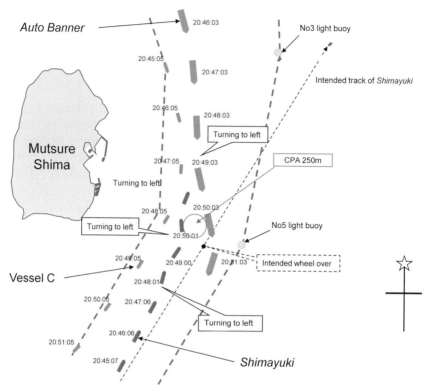

Fig. 11.3: AIS tracks of *AUTO BANNER* and *SHIMAYUKI* (Adapted from JTSB 2014e).

Color version at the end of the book

Teamwork

(1) Lookouts and a radar watchman at the bridge performed their own roles and responsibilities.

(2) The master and the chief navigator did not perform their roles, since critical information reported by crewmembers about their own ship's position and the *AUTO BANNER* was ignored or not shared.

(3) From these facts, teamwork was not functioning at the time of accident.

Management (Organizational factors)

(1) The navigation manual of the *SHIMAYUKI*, created by the Training Team Group (hereinafter referred to as "navigation manual") prescribed that, when navigating westward, the Kanmon Passage vessels must navigate 200 m inward from the outer boundary of the Kanmon Passage. Intended track and track of the *SHIMAYUKI* at the time of the accident deviated from the navigational manual.

(2) Checking the passage planning by the organization of the vessel was not conducted.

Collision between the *YONG SHENG VII* and the *HOKUEI No. 18*

The course of accident and contributing factors

Maneuvering of the *YONG SHENG VII*

The *YONG SHENG VII* left the Port of Kin-nakagusuku in Okinawa, bound for the Port of Inchon in Republic of Korea at 19:05 on 15 November 2014. The second officer (2/O), after finishing the stern station, took the operation of the engine telegraph and watches at the bridge. The *YONG SHENG VII*, while lowering both anchors near the water's surface so that they could be anchored immediately at any moment, proceeded southeast in a 13 m dredged fairway at 5 to 7 knots (Fig. 11.4).

The master of the *YONG SHENG VII* (hereinafter referred to as "master A") recognized an inbound vessel by the radar and the naked eyes in the south-southeast direction off the East breakwater and knew the ship's name by the AIS. The master A had planned to navigate in the middle of the dredged fairway but, thinking that the *HOKUEI No. 18* would navigate the same fairway, he decided to keep own ship near to the south-west boundary of buoyed waterways.

The master A saw the *HOKUEI No. 18* displaying the masthead lights and the port sidelight, navigating on the left side of dredged fairway which lay on the port side of the *HOKUEI No. 18*, and then slowly turning toward left at the bend of the fairway. The master A was still expecting that the *HOKUEI No. 18* might sooner or later start to navigate on the right side of dredged fairway and he did not slow down own ship's speed.

At 19:18, the master A found two masthead lights of the *HOKUEI No. 18* almost on the vertical line, and the *HOKUEI No. 18* was rapidly approaching own vessel. He felt the danger of collision. The master A gave five short rapid blasts on the whistle signal and five short rapid flashes on

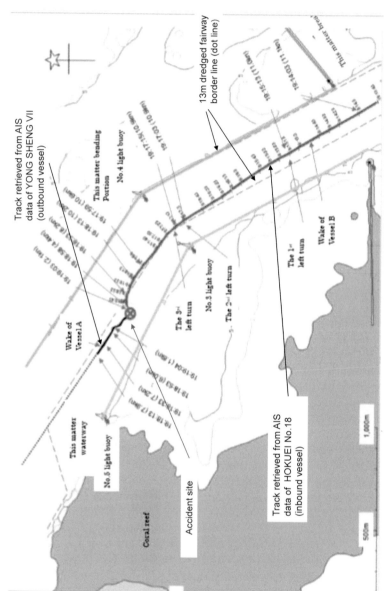

Fig. 11.4: AIS tracks of *YONG SHENG VII* and *HOKUEI No. 18* (Adapted from JTSB 2016a).

Color version at the end of the book

light signal towards the *HOKUEI No. 18* continually, with the intention that the *HOKUEI No. 18* should alter her course to starboard. At the same time, the master instructed the 2/O to call the *HOKUEI No. 18* with VHF.

There was no response from the *HOKUEI No. 18*. Facing immediate danger of collision, the master A ordered the chief officer on the bow station to let go the port anchor and the 2/O to take a full astern. Approximately 20 seconds later, the *YONG SHENG VII* collided with the midship part of the *HOKUEI No. 18*.

These are contributing factors related to the accident on the *YONG SHENG VII*.

Environment

There were shallow waters and land area surrounding 13 m dredged fairway, except for the south boundary of the fairway. With regards to the *YONG SHENG VII*, safe margin for maneuvering was constrained at the time of accident.

Condition of operators

When the *HOKUEI No. 18* slowly turned left at the bend of the dredged fairway, the master A assumed that *HOKUEI No. 18* might sooner or later start to proceed on the right side of fairway, and the master A continued to maintain own ship's speed. At that time, the distance between the two vessels was close, around 540 m to 640 m. The master A's assumption allowed him to lose an appropriate opportunity to sound warning signals and take collision avoidance actions in ample time.

Communication

After feeling the danger of collision at 19:18, the master A made the whistle signals and light signals with the intention that the *HOKUEI No. 18* would alter her course to starboard. Furthermore, he instructed the 2/O to call the *HOKUEI No. 18* on VHF. All efforts to communicate with the *HOKUEI No. 18* failed. The master A was uncertain of how to avoid collision, even before and after he felt the danger of collision, as all measures to communicate with the *HOKUEI No. 18* had failed.

Maneuvering of the *HOKUEI No. 18*

The *HOKUEI No. 18*, with the master (hereinafter referred to as "master B") and four crewmembers onboard, left the gravel excavation water off

Maejima-island, Okinawa, at 15:40 on 15 November 2014, scheduled to arrive –5.5 m quay at the Port of Kin-nakagusuku in Okinawa at 19:00. The master B came to the bridge off the water of Kudakakuchi, took the conn from the chief officer (hereinafter referred to as "C/O B").

The master B stood near the steering stand and maneuveed the vessel using the automatic steering mode as well as the radar and GPS plotter. The C/O B and an able seaman at the bow station, in preparing for entering the port, performed the task of pulling out the mooring ropes on the deck, and completed the task at 18:50. The C/O B then decided to wait at the accommodation because there was enough time left before the arrival at the quay.

The master B, in the vicinity of the East breakwater, saw the radar screen set up at 2.5 miles range and assumed that there were no outbound vessels as there was no radar image of the other vessels. The master B was in a hurry as they were about 30 minutes behind the scheduled estimated time of arrival, and proceeded to the left side of the fairway which lay on the port side of the *HOKUEI No. 18* with her speed full ahead at 11 knots after passing through the East breakwater (Fig. 11.4).

The master B recognized the white light of a vessel that looked like a fishing boat (hereinafter referred to as "Vessel C") off the starboard bow of own vessel in East waters off the bend of fairway crossing, and carefully observed the movement of Vessel C. After Vessel C had passed by own vessel, the master B altered the course to port by setting the knob of the automatic steering system to left by 10°, while looking at the No. 3 light buoy at the bend in the port side. The master B attempted to shorten the route without noticing the presence of the *YONG SHENG VII*.

At 19:18 the master B further turned the knob around 10° to the left, and, when the *HOKUEI No. 18* was about to reach the 10° turning to the port, saw the two masthead lights of the *YONG SHENG VII* lined up in the vertical direction for the first time. The distance between the two vessels was around 540 m to 640 m. The master B felt the danger of collision.

The master B switched from automatic steering mode to manual, took the helm to hard port and full astern. The *HOKUEI No. 18* collided with the *YONG SHENG VII*. Upon hearing the astern engine sound, the C/O B and the able seaman went out of the accommodation to the upper deck thinking that the vessel was approaching the quay.

These are contributing factors related to the accident on the *HOKUEI No. 18*.

Passage planning

(1) The master B planned and proceeded in the middle of lines of buoyed waterways. Rule 9 in COLREGs states that a vessel must keep as near to the outer limit of the channel or fairway which lies on her starboard side as is safe and practicable.

(2) From the fact above (1), planning stage of the passage planning was insufficient.

Rules

(1) Rule 6 in COLREGs states that every vessel shall at all times proceed at a safe speed so that she can take proper and effective action to avoid collision and be stopped within a distance appropriate to the prevailing circumstances and conditions. The *HOKUEI No. 18* continued proceeding at full ahead within the port.

(2) The *HOKUEI No. 18* proceeded in the middle of lines of buoyed waters.

(3) From those facts, the *HOKUEI No. 18* deviated from the rule at the time of accident.

Condition of operators

(1) The master B had paid attention on the Vessel C prior to the accident; therefore, it is possible that he did not recognized the movement of the *YONG SHENG VII* until just before the occurrence of collision. His situation awareness was incomplete.

(2) The workload of master B, single-handed at the bridge, might have exceeded his ability to handle all the tasks that he faced at the time of the accident, including lookouts and observing of vessel C, maneuvering the ship in restricted waterways at night, and communication. He might not have noticed the whistle and light signals made by the *YONG SHENG VII* due to distraction of attention.

Communication

(1) The master B, at the time of entry into the Port of Kin-nakagusuku, thought that the C/O B was in the bow station.

(2) The C/O B did not report to the master B when he left the bow station, as a result, the C/O B was not able to inform the master B of any safety information about the *YONG SHENG VII.*

(3) From these facts, communication among crewmembers was insufficient.

Teamwork

(1) The C/O B did not perform his role and responsibilities in the bow station because he left the station without permission from the master B. He had responsibilities to report safety information about approaching vessels to the bridge. Any critical information about the *YONG SHENG VII* was not reported to the master B.

(2) From the fact above (1), teamwork was not functioning at the time of the accident.

Management (Organizational factors)

(1) The ship owner had not supervised the safe ship operation of the *HOKUEI No. 18*, instead, the company left it to the master B.

11.5 Difference between an accident and an incident

11.5.1 Collisions

In case of the *AUTO BANNER* and *SHIMAYUKI* incident, the *SHIMAYUKI* was heading to an accident trajectory. This is supported by the analysis of contributing factors on the *SHIMAYUKI* that passage planning, condition of operators, environment, communication, teamwork and management were insufficient at the time of accident. However, the *AUTO BANNER* managed to prevent the collision with the *SHIMAYUKI* by taking her helm to port and sounding two short blasts towards the *SHIMAYUKI* continuously under special circumstances and immediate danger. The master of the *SHIMAYUKI* stated that the two short blasts sounded by the *AUTO BANNER* convinced him of the movement of the *AUTO BANNER*. Both masters were certain of how to avoid a collision. Another important aspect of preventing the collision is that two vessels made good use of whistle signals, not using VHF, and both masters were certain of how to avoid collision.

On the other hand, in the case of the *YONG SHENG VII* and *HOKUEI No. 18* accident, the *YONG SHENG VII* took precautionary measures to prevent a collision by proceeding to the far-right side of the fairway. However, both sound and light signals made by the *YONG SHENG VII* were not recognized by the master of the *HOKUEI No. 18* who did not notice the presence of the *YONG SHENG VII* until the accident occurred.

In other cases of collisions, the same thing that one vessel did not or could not identify the approaching vessel is applied, e.g., the collision between the *DAIO DISCOVERY* and the *AURORA SAPPHIRE* (Chapter 4) and the collision between the *MEDEA* and the *KOSEI-MARU* (Chapter 8).

In the case of the collision between the *KOTA DUTA* and the *TANYA KARPINSKAYA* (Chapter 4), the detailed analysis based on VDR readout clarifies that VHF communication on conduct of vessel gave both masters false confidence on how to avoid collision by agreeing movement of the vessels that often deviated from CORLEGs. As a result, they lost limited time to avoid collision under immediate and danger situation.

11.5.2 Occupational casualties

In the case of the *KYOKUHOU-MARU No. 2* (Chapter 4), a worker hearing the occurrence of an accident went to the pier and saw three crewmembers wrapping a rope around their waists and preparing to enter the cargo tank. The worker realized a secondary disaster would happen because he had learned of similar cases during education and training sessions conducted by his company. The worker intervened and recommended crewmembers not to enter the cargo tank until ambulance arrival. The secondary disaster was prevented. When regarding the three crewmembers as operators, the event that they were about to enter the cargo tank was an incident.

In the case of the cargo vessel *SINGAPORE GRACE* (Chapter 4), the foreman, who was informed that Driver B had collapsed, entered the cargo hold No. 3 to rescue Driver B together with Operator C and Operator F without measuring the atmosphere in the cargo hold No. 3. As a result, the foreman inhaled oxygen-deficient air and developed anoxia. The foreman was not aware of oxygen-deficient atmosphere in cargo hold No. 3. Operator C, together with Operator F, then entered cargo hold No. 3 wearing gas masks in order to rescue the foreman and Driver B, inhaled oxygen-deficient air and developed anoxia. There was no external third party who knew the danger of entering the enclose space without measuring the atmosphere and could prevent those workers from entering the cargo hold.

In cases of occupational casualties, especially during entering an enclose space, considering these cases, when an operator is heading to accident trajectory without measuring the atmosphere, an external third party who knows how to avoid an accident by lessons learned can prevent the operator from getting involved in the accident, and then the event becomes an incident. If there is no said external third party around the site, an accident occurs.

11.5.3 New model showing accident and incident mechanism

These studies provide a new model on accident and incident occurrence mechanism. Risk controls that are reined in by an external third party are related to stopping the progress to an accident, resulting in an incident. Figure 11.5 shows the new incident occurrence mechanism adding up to the original model in Fig. 11.2.

Risk controls by the external third party means that, in the collision between the *AUTO BANNER* and the *SHIMAYUKI*, the pilot of the *AUTO BANNER* ordered to take her helm to port and sound two short blasts towards the *SHIMAYUKI* continuously under special circumstances and immediate danger, resulting in convincing the master of the *SHIMAYUKI* of the movement of the *AUTO BANNER* and of following the requested

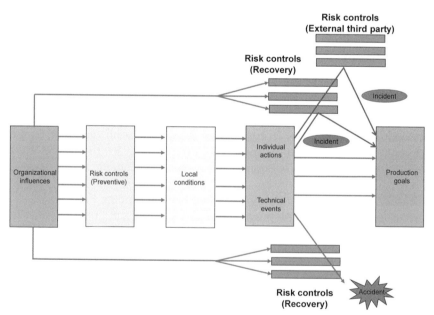

Fig. 11.5: New accident and incident occurrence mechanism.

maneuvering. On the part of the *SHIMAYUKI*, all risk controls or defensive layers had been penetrated by the accident trajectory at the time of incident.

The same thing is applied to an incident of occupational casualties. In cases of enclosed space entry that were explained in the chemical tanker *KYOKUHOU-MARU No. 2* and the cargo vessel *SINGAPORE GRACE*, even though all the defensive layers on the three crewmembers of the *KYOKUHOU-MARU No. 2* had been penetrated by the accident trajectory, progress to the accident was prevented by the external third party who knew how to avoid an accident due to education and training.

Therefore, new definition on the incident occurrence mechanism is that an incident occurs either when part of defensive layers has worked, or even if defensive layers have been penetrated by the accident trajectory an external third party who knows how to avoid accident gets involved in the progress of the accident and succeeds in functioning as an additional defensive layer to the operator concerned.

11.6 Conclusions

As an accident investigator, the most frequent questions received from the shipping industry are how to prevent the accidents. This book and this chapter are constructed to answer those questions. This chapter illustrates the systematic accident prevention which incorporates and combines risk management, principles of quality management, accident model, and statistics into one system in order for readers to comprehend and execute the systematic methods that fit to their own companies. The systematic accident prevention can be applicable to other field if the accident model is modified to fit them and be applicable to incidents as well.

References

Anca, J.M. 2007. Multimodal Safety Management and Human Factors: Crossing the Borders of Medical, Aviation, Road and Rail Industries. Ashgate Publishing Limited, Aldershot, UK.

Adams, M.R. 2006. Shipboard Bridge Resource Management. Nor'easter Press, Eastport, USA.

[ATSB] Australian Transport Safety Bureau. 2001. Marine Safety Investigation Report 162: Independent investigation into the grounding of the Malaysian flag container ship *Bunga Teratai Satu*.

[ATSB] Australian Transport Safety Bureau. 2008. ATSB Transport Safety Research Report: Aviation Research and Analysis Report-AR-2007-053, Analysis, Casualty and Proof in Safety Investigations.

Baltes, P.B. and Lindenberger, U. 1997. Emergence of powerful connection between sensory and cognitive functions across the adult life span: A new window to the study of cognitive aging? Psychol Aging 12: 12–21.

Caldwell, J.A. and Caldwell, J.L. 2003. Fatigue in Aviation: A Guide to Staying Awake at the Stick. Ashgate, Surrey, UK.

Chauvin, C., Lardjane, S., Morela, G., Clostermann, J. and Langard, B. 2013. Human and organizational factors in maritime accidents: Analysis of collisions at sea using the HFACS. Accid Anal Prev 59: 26–37.

Cockcroft, A.N. and Lameijer, J.N.F. 1982. A guide to the collision avoidance rules. Stanford Maritime Limited, London, UK.

Dekker, S. 2006. The Field Guide to Understanding Human Error. Ashgate Publishing Limited, Surrey, UK.

[DOE] Department of Energy. 1999. Conducting Accident Investigations: DOD Workbook.

[DOT] Department of Transport, Her Majesty's Stationery Office. 1987. MV Herald of Free Enterprise Report of Court No. 8074 Formal Investigation.

DNV-GL, Gard and The Swedish Club. 2016. Anchor loss-technical and operational challenges and recommendations.

Endsley, M.R. 2000. Theoretical understandings of situation awareness: a critical review. pp. 3–32. *In*: Endsley, M.R. and Garland, D.J. [eds.]. Situation Awareness Analysis and Measurement. Lawrence Erlbaum Associates, Inc., Mahwah, USA.

Fisher, R.P. and Geiselman, R.E. 1992. Memory-enhancing Techniques for Investigative Interviewing: The Cognitive Interview. Charles C. Thomas Publisher, Springfield, USA.

Fukuoka, K. 2015. Visualization of holes and relationships between holes and latent conditions. pp. 215–221. *In*: Weintrit, A. [ed.]. Activities in Navigation: Marine Navigation and Safety of Sea Transportation. CRC Press, London, UK.

Fukuoka, K. 2016a. Visualization of a hole and accident preventive measures based on the Swiss cheese model developed by risk management and process approach. WMU Journal of Maritime Affairs 15: 127–142.

Fukuoka, K. 2016b. Model of marine accidents developed by the Swiss cheese model and recommendations for systematic accident prevention. PhD thesis, Kobe University, Kobe, Japan (Japanese).

Fukuoka, K. and Furusho, M. 2016a. Relationship between latent conditions and the characteristics of holes in marine accidents based on the Swiss cheese model. WMU Journal of Maritime Affairs 15: 267–292.

Fukuoka, K. and Furusho, M. 2016b. Influence of human factors on the occurrence of accidents-quantification of the influence of SHEL elements and latent conditions. Journal of Maritime Researches 6: 1–8.

Fukuoka, K. 2017. Accident models and their applicable industries. Journal of Science of Labour 93.2: 48–60 (Japanese).

Fukuoka, K. and Furusho, M. 2017. Convergence of the epidemiological and systemic accident models. Journal of Maritime Researches 7: 1–11.

[GL] Germanischer Lloyd. 2011. GL Academy: Oil and Chemical Tanker: Technical and Operational Aspects.

Grandjean, E. 1969. Fitting the Task to the Man; An Ergonomic Approach. Taylor & Francis, London, UK.

Grech, M.R., Horberry, T.J. and Koester, T. 2008. Human Factors in the Maritime Domain. CRC Press Taylor & Francis Group, Boca Raton, USA.

Hawkins, F.H. 1987. Human Factors in Flight. Gower Technical Press, Aldershot, UK.

Heinrich, H.W. 1959. Industrial Accident Prevention. McGraw-Hill Book Company, Inc., New York, USA.

Hollnagel, E. 1998. Cognitive Reliability and Error Analysis Method: CREAM. Elsevier Science Ltd., Oxford, UK.

Hollnagel, E. 2004. Barriers and Accident Prevention. Ashgate Publishing Limited, Aldershot, UK.

Hollnagel, E. and Speziali, J. 2008. Study on Developments on Accident Investigation Methods: A Survey of the "State-of-the-Art". SKI Report 2008:50. Swedish Nuclear Power Inspectorate, Stockholm, Sweden.

Holmes, T.H. and Rahe, T.H. 1967. The social readjustment rating scale. J Psychosom Res 11: 213.

[IACS] International Association of Classification Societies. 2016. Requirements Concerning Mooring, Anchoring and Towing. IACS Req. 1981/Rev.6 2016/Corr.2.

[ICAO] International Civil Aviation Organization. 1993. Human Factors Digest No. 7: Investigation of human factors in accidents and incidents, Circular 240-AN/144.

[ICAO] International Civil Aviation Organization. 2001. Manual of Aircraft Accident and Incident Investigation: Part 1 Organization and Planning.

[IEC] International Electrotechnical Commission. 2009. IEC/ISO 31010: 2009 Risk Management-Risk Assessment Techniques.

Iizuka, Y. 2009. Gendai Hinshitsu Kanri Souron. Asakura Publishing Company, Tokyo, Japan (Japanese).

[IMO] International Maritime Organization. 1993. International Management Code for the Safe Operation of Ships and for Pollution Prevention (International Safety Management (ISM) Code). Resolution A 741(18).

[IMO] International Maritime Organization. 2000a. Guidelines on Ergonomic Criteria for Bridge Equipment and Layout. MSC/Circ.982.

[IMO] International Maritime Organization. 2000b. Guidelines for Voyage Planning. Resolution A 893(21).

[IMO] International Maritime Organization. 2000c. Amendments to the Code for the Investigation of Marine Casualties and Incidents: The IMO/ILO Process for Investigating Human Factors. Resolution A 884(21).

[IMO] International Maritime Organization. 2001. Guidance on Fatigue Mitigation and Management. MSC/Circ.1014.

[IMO] International Maritime Organization. 2002. Guidelines on Voyage Data Recorder (VDR) Ownership and Recovery. MSC/Circ.1024.

[IMO] International Maritime Organization. 2003. Role of the Human Element. MSC 77/17.

[IMO] International Maritime Organization. 2008. Code of the International Standards and Recommended Practices for a Safety Investigation into a Marine Casualty or Marine Incident (Casualty Investigation Code). Resolution MSC.255(84).

[IMO] International Maritime Organization. 2011. Revised Recommendations for Entering Enclosed Spaces Aboard Ships. Resolution A 1050(27).

[IMO] International Maritime Organization. 2012. Adoption of Revised Performance Standards for Shipborne Voyage Data Recorders (VDRs). Resolution MSC.333(90).

[IMO] International Maritime Organization. 2013. Safe Mooring: A Guide to Prevent Accidents While Mooring. MSC 92 INF.11.

[IMO] International Maritime Organization. 2014a. Guidelines to Assist Investigators in the Implementation of the Casualty Investigation Code (Resolution MSC.255(84)). Resolution A 1075(28).

[IMO] International Maritime Organization. 2014b. Model Course 3.11: Safety Investigation into Marine Casualties and Marine Incidents. IMO, London, UK.

[ISO] International Organization of Standardization. 2008a. ISO 9000 Introduction and support package: guidance on the concept and use of the process approach for management systems. ISO/TC 176/SC 2/N 544R3.

[ISO] International Organization of Standardization. 2008b. ISO 9001: 2008 Quality Management Systems-Requirements.

[ISO] International Organization of Standardization. 2009. ISO 31000: 2009 Risk Management-Principles and Guidelines.

[ISO/IEC] International Organization of Standardization and International Electrotechnical Commission. 1999. ISO/IEC Guide 51: 1999(E) Safety Aspects-Guidelines for their Inclusion in Standards.

Italian Ministry of Infrastructure and Transport. 2012. Marine Accident Investigation C/S *Costa Concordia* 13 January 2012. MSC90 of IMO.

[JCG] Japan Coast Guard. 2015. Marine accident statistics and prevention in 2014 (Japanese).

Japan Federation of Pilots' Associations. 2012. Pilot Transfer Arrangements and their Operations. Japan Federation of Pilots' Associations, Tokyo, Japan (Japanese).

[JTSB] Japan Transport Safety Board. 2011a. Marine accident investigation report: Collision between cargo vessel *DAIO DISCOVERY* and *AURORA SAPPHIRE*, MA2011-5-7 (Japanese).

[JTSB] Japan Transport Safety Board. 2011b. Marine accident investigation report: Fatality of a crewmember onboard chemical tanker *KYOKUHOU-MARU No. 2*, MA2011-7 (Japanese).

[JTSB] Japan Transport Safety Board. 2011c. Marine accident investigation report: Grounding of cargo vessel *LANA*, MA2011-6-17 (Japanese).

[JTSB] Japan Transport Safety Board. 2011d. Marine accident investigation report: Fire on vehicle carrier *PYXIS*, MA2011-10.

[JTSB] Japan Transport Safety Board. 2012a. Marine accident investigation report: Collision between cargo vessel *MEDEA* and fishing vessel *KOSEI-MARU*, MA2012-9-2 (Japanese).

[JTSB] Japan Transport Safety Board. 2012b. Marine accident investigation report: Fatality of workers onboard cargo vessel *SINGAPORE GRACE*, MA2012-4.

[JTSB] Japan Transport Safety Board. 2012c. Annual report 2012.

[JTSB] Japan Transport Safety Board. 2013. Annual report 2013.

[JTSB] Japan Transport Safety Board. 2014a. Marine accident investigation report: Contact of sea wall, container vessel *FLEVODIJK*, MA2014-1.

[JTSB] Japan Transport Safety Board. 2014b. Marine accident investigation report: Collision between container vessel *Kota Duta* and cargo vessel *Tanya Karpinskaya*, MA2014-5.

[JTSB] Japan Transport Safety Board. 2014c. Marine accident investigation report: Collision between bulk carrier vessel *NIKKEI TIGER* and fishing vessel *HORIEI-MARU*, MA2014-6.

[JTSB] Japan Transport Safety Board. 2014d. Marine accident investigation report: Fatalities of crewmembers of container vessel ANNA MAERSK, MA2014-2.

[JTSB] Japan Transport Safety Board. 2014e. Marine accident investigation report: Collision between car carrier *AUTO BANNER* and training vessel *Shimayuki*, MA2014-10-1.

[JTSB] Japan Transport Safety Board. 2014f. Annual report 2014.

[JTSB] Japan Transport Safety Board. 2015a. Marine accident investigation report: Fatality of a crewmember of cargo vessel *ONOE*, MA2015-7.

[JTSB] Japan Transport Safety Board. 2015b. Annual report 2015.

[JTSB] Japan Transport Safety Board. 2016a. Marine accident investigation report: Collision between cargo vessel *YONG SHENG VII* and dredger carrier *HOKUEI No. 18*, MA2016-8.

[JTSB] Japan Transport Safety Board. 2016b. Marine accident investigation report: Collision between cargo vessel *BEAGLE III* and container vessel *PEGASUS PRIME*, MA2016-5.

[JTSB] Japan Transport Safety Board. 2017a. Annual Report 2017.

[JTSB] Japan Transport Safety Board. 2017b. Marine accident investigation report: Grounding of cargo vessel *CITY*, MA2017-9.

Leveson, N.G. 2011. Engineering a Safer World: Systems Thinking Applied to Safety. The MIT Press, London, UK.

Lu, C.S., Lai, K.H., Lun, Y.H. and Cheng, T.C. 2012. Effects of national culture on human failures in container shipping: the moderating role of Confucian dynamism. Accd Anal Prev 49: 457–469.

Macrae, C. 2009. Human factors at sea: Common patterns of error in groundings and collisions. Marit Policy Manag 36: 21–38.

Maguire, R. 2006. Safety Cases and Safety Reports: Meaning, Motivation and Management. Ashgate Publishing Limited, Surrey, UK.

[MAIA] Marine Accident Inquiry Agency. 2003. Grounding of cargo vessel *COPE VENTURE* (Japanese).

[MAIA] Marine Accident Inquiry Agency. 2004. Accident caused by sleepiness of single navigation watcher at Inland waters (Japanese).

[MAIA] Marine Accident Inquiry Agency. 2006. Marine accident analysis No. 6: Typhoon and accidents (Japanese).

[MAIA] Marine Accident Inquiry Agency. 2007. Marine accident analysis No. 7: Accidents in restricted visibility/fog (Japanese).

[MAIB] Marine Accident Investigation Branch. 2004. Bridge Watchkeeping Safety Study.

[MAIB] Marine Accident Investigation Branch. 2007. Fishing accident flyer.

[MAIB] Marine Accident Investigation Branch. 2008a. Accident Report: Report on the investigation of the grounding of *CFL Performer* Haisborough Sand, North Sea, 12 May 2008 (Report No. 21/2008 December 2008).

[MAIB] Marine Accident Investigation Branch. 2008b. Accident Report: Report on the investigation of *Sichem Melbourne* making heavy contact with mooring structures at Coryton Oil Refinery Terminal on 25 February 2008 (Report No. 18/2008 October 2008).

[MAIB] Marine Accident Investigation Branch. 2012. Accident Report: Grounding of *CSL THAMES* in the Sound of Mull, 9 August 2011 (Report No. 2/2012 March 2012).

[MAIB] Marine Accident Investigation Branch. 2014a. Accident Report: Report on the investigation of the grounding of *Ovit* in the Dover Strait, 18 September 2013 (Report No. 24/2014 September 2014).

[MAIB] Marine Accident Investigation Branch. 2014b. Accident Report: Report on the investigation of the collision between *CMA CGM Florida* and *Chou Shan*, 140 miles east of Shanghai, East China Sea on 19 March 2013 (Report No. 11/2014 May 2014).

[MAIB] Marine Accident Investigation Branch. 2016. Accident Report: Report on the investigation of the capsize and sinking of the cement carrier *Cemfjord* in the Pentland Firth, Scotland with the loss of all eight crew on 2 and 3 January 2015 (Report No. 8/2016 April 2016).

[MAIB] Marine Accident Investigation Branch. 2017a. Accident Report: Report on the investigation of the grounding of *Muros* Haisborough Sand, North Sea, 3 December 2016 (Report No. 22/2017 October 2017).

[MAIB] Marine Accident Investigation Branch. 2017b. Accident Report: Report on the investigation of the failure of a mooring line on board the LNG carrier *Zarga* while alongside the South Hook Liquefied Natural Gas terminal, Milford Haven (Report No. 13/2017 June 2017).

[MCA] Maritime and Coastguard Agency. 2006. Marine Guidance Note MGN 324 (M+F), Radio: Operational Guidance on the Use of VHF Radio and Automatic Identification Systems (AIS) at Sea.

Milne, R. and Bull, R. 1999. Investigative Interviewing: Psychology and Practice. John Wiley & Sons, Ltd., Chichester, UK.

Moore-Ede, M. 1993. The 24-Hour Society: The Risks, Costs and Challenges of a World that Never Stops. Judy Piatkus Ltd., London, UK.

[NTSB] National Transportation Safety Board. 1990. Marine Accident Report: Grounding of the U.S. tankship *EXXON VALDEZ* on Bligh Reef, Prince William Sound near Valdez, Alaska. March 24, 1989 (Report No. NTSB/MAR-90/04).

Nishino, S. 2017. The Stanford Method for Ultimate Sound Sleep. Sanmark Publishing, Tokyo, Japan (Japanese).

Norman, D.A. 1988. The Psychology of Everyday Things. Basic Books, New York, USA.

OHSAS Project Group 2007. 2007. OHSAS 18001: 2007: Occupational health and safety management systems—Requirements.

Parrott, D.S. 2011. Bridge Resource Management for Small Ships: The Watchkeeper's Manual for Limited-Tonnage Vessels. The McGraw-Hill Companies, Blacklick, USA.

Perrow, C. 1999. Normal Accident. Princeton University Press, New Jersey, USA.

Rasmussen, J., Duncan, K. and Leplat, J. 1987. New Technology and Human Error. John Wiley & Sons Ltd., New York, USA.

Rasmussen, J. 1997. Risk management in a dynamic society: A modelling problem. Saf Sci 27: 183–213.

Reason, J. 1990. Human Error. Cambridge University Press, New York, USA.

Reason, J. 1997. Managing the Risks of Organizational Accidents. Ashgate Publishing Limited, Surrey, UK.

Reason, J. and Hobbs, A. 2003. Managing Maintenance Error: A Practical Guide. Ashgate Publishing Limited, Surrey, UK.

Reason, J., Hollnagel, E. and Paries, J. 2006. Revisiting the Swiss cheese model of accidents. Eurocontrol Experimental Center, EEC Note No. 13/06. France.

Reason, J. 2008. The Human Contribution: Unsafe Acts, Accidents and Heroic Recoveries. Ashgate Publishing Limited, Surrey, UK.

Schröder-Hinrichs, J.-U., Baldauf, M. and Ghirxi, K.-T. 2011. Accident investigation reporting deficiencies related to organizational factors in machinery space fires and explosions. Accid Anal Prev 43: 1187–1196.

Strauch, B. 2004. Investigating Human Error: Incidents, Accidents, and Complex Systems. Ashgate Publishing Limited, Surrey, UK.

Swift, A.J. 2000. Bridge Team Management: A Practical Guide. The Nautical Institute, London, UK.

Underwood, P. and Waterson, P. 2014. System thinking, the Swiss Cheese Model and accident analysis: A comparative systemic analysis of the Grayrigg train derailment using the ATSB, AcciMap and STAMP models. Accid Anal Prev 68: 75–94.

US Geological Survey. 2014. 2013 Update on Sea Otter Studies to Assess Recovery from the 1989 *Exxon Valdez* Oil Spill, Prince William Sound, Alaska (Open-File Report 2014–1030).

Wiegmann, D. and Shappell, S. 2003. A Human Error Approach to Aviation Accident Analysis: The Human Factors Analysis and Classification. Ashgate Publishing Limited, Surrey, UK.

Woods, D.D. 1984. Some results on operator performance in emergency events. Institute of Chemical Engineers Symposium Ser 90: 21–31.

Yokomizo, Y. and Komatsubara, A. 2013. Ergonomics or Engineers. Japan Publication Service, Tokyo, Japan (Japanese).

Index

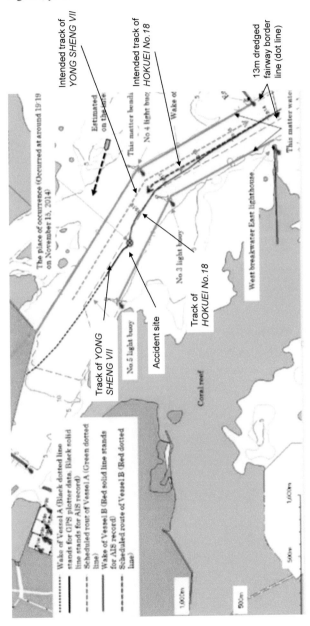

Chapter 4

Fig. 4.11, p. 79

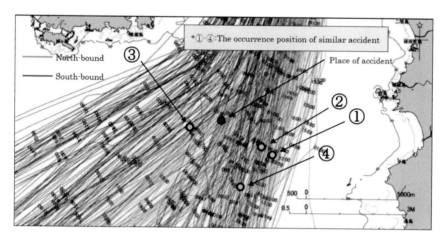

Chapter 8

Fig. 8.12, p. 177

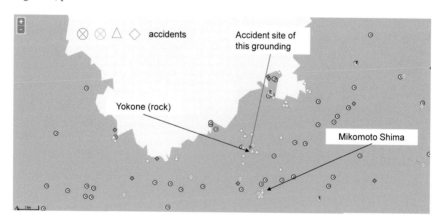

266

Chapter 11
Fig. 11.3, p. 240

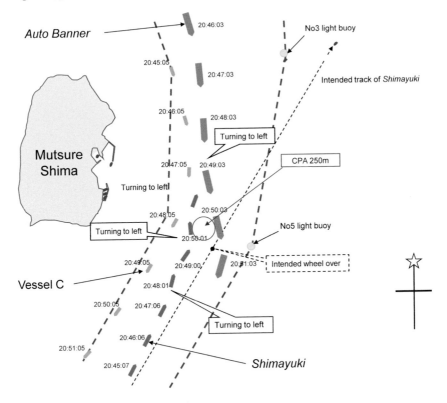

Auto Banner

No3 light buoy

20:46:03

20:45:05

20:47:03

Intended track of *Shimayuki*

20:46:05

20:48:03

Turning to left

Mutsure
Shima

CPA 250m

20:47:05   20:49:03

Turning to left

20:48:05   20:50:03

Turning to left

No5 light buoy

20:50:01

20:49:05   20:49:00   20:51:03   Intended wheel over

Vessel C

20:48:01

20:50:05   20:47:06

Turning to left

20:46:06

20:51:05

20:45:07   *Shimayuki*

Chapter 11

Fig. 11.4, p. 242

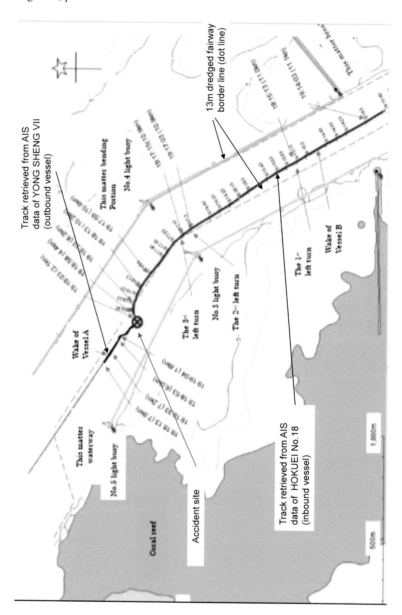